T0244126

THIS VOLCANIC ISLE

THIS VOLCANIC ISLE

THE VIOLENT PROCESSES THAT FORGED THE BRITISH LANDSCAPE

ROBERT MUIR-WOOD

OXFORD
UNIVERSITY PRESS

Great Clarendon Street, Oxford, OX2 6DP,
United Kingdom

Oxford University Press is a department of the University of Oxford.
It furthers the University's objective of excellence in research, scholarship,
and education by publishing worldwide. Oxford is a registered trade mark of
Oxford University Press in the UK and in certain other countries

Published in the United States of America by Oxford University Press
198 Madison Avenue, New York, NY 10016, United States of America

British Library Cataloguing in Publication Data
Data available

Library of Congress Control Number: 2023923234

ISBN 978–0–19–887162–0

Printed and bound in the UK by
Clays Ltd, Elcograf S.p.A.

Acknowledgements

I pursued the original research for *This Volcanic Isle* thirty years ago.[1] The admitted motive was to help explain where future earthquakes might rupture. For me there was another reason: to retrieve an untold, deep-time saga.

I then moved on to research other worlds: the science underlying the risk modelling of climate and terrestrial catastrophes.[2] Meanwhile I sustained the vision, one day, to return and write this book. This foundational history for the land and subsea surrounds of Britain and Ireland, I firmly believed, deserved a broad readership.

My research had been commissioned by David Mallard (then at the Central Electricity Generating Board), who deserves the most fulsome acknowledgement. I would also like to thank colleagues from this time: (†)Bryan Skipp, (†)Ian Higginbottom, Charles Melville, Gordon Woo, Willie Aspinall, John Gutmanis, Keith Sizer, and Robert Maddock.

The research spilled over into projects on active tectonics offshore Norway (thanks to Hilmar Bungum and Tor Loken) and in mainland Sweden (thanks to Martin Ekman, Lars Ericsson and the most accomplished field geologist I have ever met: (†)Robert Lagerbäck). I learnt where to expect future volcanic eruptions in Iceland with Scott Steedman and Pall Einarsson.

1. R. Muir Wood (1989). 'Fifty million years of "passive margin" deformation in North West Europe.' In: *Earthquakes at North-Atlantic Passive Margins: Neotectonics and Postglacial Rebound*, NATO ASI Series, vol. 266, Dordrecht: Springer, pp. 7–36, https://doi.org/10.1007/978-94-009-2311-9_2.
2. See the 2016 book *The Cure for Catastrophe: How We Can Stop Manufacturing Natural Disasters*, published by Oneworld (UK) and Basic Books (US).

Much of the evidence that underlies this historic reconstruction is reported in short (and often obscure, scarcely cited) scientific papers. Many of these were generated from the detailed mapping of the British Geological Survey and I recall animated discussions with Robin Wingfield and Dave Tappin. Others were scientifically generous by-products of commercial oil and gas exploration. A few were from reports of academic field geology. I have visited many of the 'sacred' tectonic sites but my contribution has been to 'bake a cake' out of others' ingredients.

I thank Geof King for my introduction to earthquake geology, Claudio Vita Finzi for stylish neotectonics, Nigel Harris for field work in the magnificent Pyrenees on the back of his Triumph 500, brother (†)Paul Muir Wood who read and commented on an early draft, the metaphorical Miles Warde, and above all my wife Elizabeth, walking with me in many of the landscapes in this story.

I am pleased, finally, to bring this foundational history home. I would especially like to thank Latha Menon at OUP who championed this project through her own geological enthusiasms and Jamie Mortimer who patiently shepherded the editing along with the production of many new figures. This book is dedicated to my children, who tolerated unexplained geological diversions on their holidays and have still gained a love for the mountains.

Contents

List of Illustrations

Frontispiece

Chapter 1

Chapter 2

Chapter 3

Chapter 4

Chapter 5

Chapter 6

Chapter 7

Chapter 8

Chapter 9

Chapter 10

Chapter 11

Chapter 12

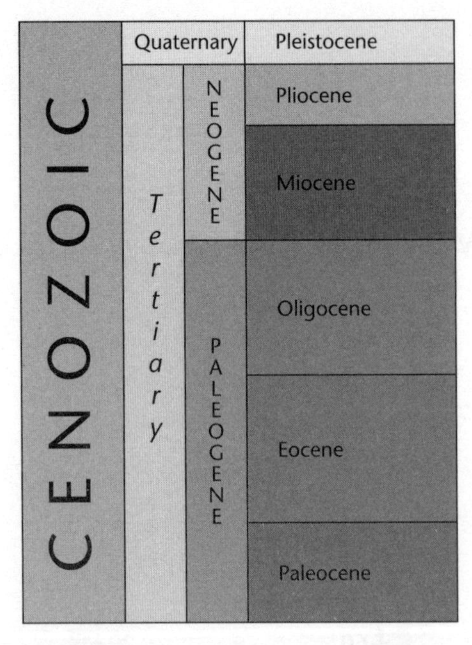

Figure 1. Subdivisions of the Cenozoic

I

Voyagers in deep time

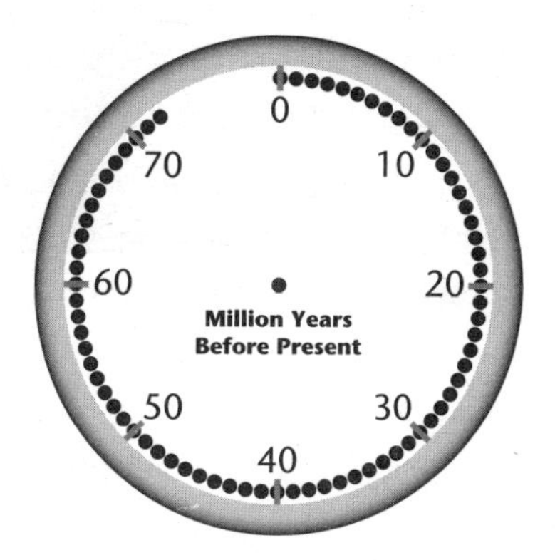

Dusk on a clear summer evening. We are sitting at the crest of the northern scarp of the Berkshire Downs. A warm breeze ripples the grass. The sky overhead is cloudless. Far below, a train, like a toy, is silently moving across the plain, passing a cricket pitch next to the church. Beyond the mosaic of fields and woods the distant Cotswold Hills are tinged with mauve.

Perhaps the view draws us into talking of the past, of the people who first cut down the forest, burnt the stumps, and ploughed the soil; and whose land-shaping has become our inheritance: a cantering horse gracefully carved into the steep grass slope, only to be seen from the vale below; a long barrow tomb, guarded by great 'sarsen' stones, in a beech tree stand.

Then there is an older history, of the origins of this land, with its cliffs and valleys, hills and beaches, islands and rivers, about which we seem to have few stories to tell.

Below us, on the steep grassy slope, the horse-artist's cut turf exposes white chalk rock, consolidated out of a seeming infinity of microscopic calcite discs, from the plankton that lived in the warm Cretaceous seas that once covered this region. The sea has long receded and what was once the sea floor has been raised into hills almost 300 metres high.

Lie back, rest your head on the grass, and look up while the last light fades in the north-west. Bats are flitting through the dark. Watch the stars appear, one by one, and then in a spray of pinprick lights, as the darkness intensifies. Our perspective has dramatically expanded, from the condensation trails 10,000 metres high and the far horizon 30 kilometres to the north, to the immensity of outer space.

The nearest star is 40 trillion kilometres away. That is forty followed by twelve zeroes—almost 300,000 times the distance of the Earth from the Sun. We can measure this distance by parallax: seeing how each star's location in the night sky shifts as the Earth circles the Sun. Its light takes more than four years to reach us. But other stars are a hundred—even a thousand—times farther away. Unfathomable, incredible distances.

At the centre of the cricket ground below, between the stumps, the pitch is 20 metres long. Imagine the distance between the Earth and the Sun was reduced to the length of that pitch. For scale the Sun would be the size of a honeydew melon—a little smaller than a football. At the other end of the pitch the Earth would be the diameter of a millet seed. And the nearest star? It would be almost 6,000 kilometres away—as far as Chicago.

We are going to explore the past on a timescale as far beyond our everyday experience of time as the distance to the stars exceeds the height of this White Horse hill. In this book I will talk of 'millions of years' of time just as astronomers report distances in 'light years', but we should not kid ourselves: these terms remain beyond simple comprehension.

In writing about his continent-wide travels with geologists, the American writer John McPhee called the immensity of prehistory 'deep time'. We can see the stars that manifest deep space. Now we can emulate that same shift of perspective to reveal deep time.

Only through the time-travel of storytelling can we attempt to recreate former worlds. Over millions of years, these Berkshire Downs have dissolved away, 'melted' like glacier ice. The chalk scarp was once situated 300 kilometres farther to the north-west. After vast forces raised the land, the thick layer of chalk has slowly peeled back, like the skin of an onion, not just once, but several times.

> The hills are shadows, and they flow
> From form to form, and nothing stands;
> They melt like mist, the solid lands,
> Like clouds they shape themselves and go.
>
> Tennyson, 'In Memoriam', canto 123

As the chalk seas receded, volcanoes built a chain of isolated mountains down western Britain, each perhaps 2 kilometres high. These volcanic piles have since been eroded right back to their roots. After the volcanoes, faults have interlaced these islands, spawning great earthquakes.

We live with the comfortable illusion that Britain and Ireland are stable and immutable, and yet the landscape is rich in tectonic signs.

★

The word 'tectonics' is from the Greek for a carpenter or builder. The root *tek* means 'to make'. *Tekton* is the maker—the fabricator. Tectonics then becomes 'construction'.

The word 'architect' builds on this word root. 'Archi' is the head, or the chief, so in its original meaning the architect is the 'top dog' builder, the one given the task to design the form of the building and how it will be assembled, who guides the work to cut, shape, and assemble the joists and beams.

At the end of the nineteenth century geologists borrowed the word 'tectonics' to mean the raising of mountains and the construction of landscape. Tectonics is a resonant word for these most powerful processes. The word became so embedded in its new geological meaning that tectonics was no longer used to denote 'building'.

Since the 1960s we have had the theory of 'plate tectonics' in which the outer shell of the Earth comprises a dozen rigid plates, thousands of kilometres across. The most dynamic parts of geology—mountain building, rifting, volcanoes, and earthquakes—happen principally at the plate boundaries.

Plate tectonics helps explain why there is consistency in tectonic action across a wide region. The forces that drive tectonics operate on a grand scale. By applying the rules of plate tectonics we can reconstruct a broader picture, not just isolated local instances of prehistoric earthquakes or eruptions.

Blending the words for architect and tectonics, 'architectonics' originally meant systematizing different styles of architecture. Across the earth sciences one might say that plate tectonics provides the architectonics to all the incidents of volcanic activity, active fault systems, and mountain building.

To reconstruct the tectonic story of Britain I will weave in the role of the scientist sleuths—the grand vizier 'architectonicians' like Robert Hooke, William Smith, and Charles Lyell—who unearthed the geological clues and contemplated the vast aeons of time. In the mid-twentieth century we have the geoscientists who cracked the plate tectonic code. Whether through all-male clubs like the Geological Society of London, which first admitted women fellows in 1919, or the travails of field work, women are under-represented. My own introduction to geology came through a great-aunt palaeontologist: Helen Marguerite Muir Wood, the first woman deputy director of London's Natural History Museum. I have her handwritten notes from a 1917 student geological field trip to Bovey Tracey in Devon, on the Sticklepath Fault.

And in diversions off the tectonic 'highway', we will explore some of the amazing resources created around Britain through the latest geological periods. The monster sarsen slabs that float on the landscape, and which determined the design of Stonehenge. The Olympic curling stones, all sourced from an uninhabited Scottish island. Fingal's Cave and the Giant's Causeway have been celebrated tourist destinations for more than 200 years, on account of their spontaneous geometric architecture. The Cumbrian fells, the wild Hebridean islands, the slate mountains of Wales—none of this topography existed 66 million years ago (or 66 Ma, an abbreviation for 'mega-annum'). As much upland as survives, far more has already vanished beneath the waves, including a chalk ridge running out west through the middle of the Channel, a Scottish landmass that included all the residual Orkneys and Shetlands, and the whole island subcontinent of Rockall, 400 kilometres west of mainland Scotland.

When imagining the home of the hobbits, with their doors in the steep valley sides opening into rooms cut in the soft earth, Tolkien had in mind the hills and valleys of southern England. Tolkien's evocation of the 'Shire' is based on holidays he took as a boy, to Devon and Dorset, and later where he lived in Worcestershire and Oxfordshire. But when Tolkien's 'Shire' was filmed, the crew travelled to New Zealand to find rolling soft-strata landscapes that were empty of urban sprawl and intensive farming. New Zealand is on a plate boundary and has big earthquakes tearing at the landscape along with smoking volcanoes. So how come such a dynamic landscape can masquerade for somewhere—as we were all led to believe—tectonically 'inert' or even dead? The answer? Britain is by no means tectonically lifeless. The hills of Dorset are as young as New Zealand's Bay of Islands. And not so long ago, a single great volcano marked the entrance to the Bristol Channel.

This is the story of that underlying tectonic history, starting from when the whole region lay beneath the chalk sea, 66 million years ago, and running through to the present day. It is a story that has never previously been told as a single saga, a story that explains the mountains, hills, and islands of the British Isles.

I wrote a first version of this prehistory thirty years ago, in work to explore the current and future hazard of earthquakes.[1] Unless we can identify the tectonics of the recent geological past, I argued, how will we know whether one or another fault rupture and accompanying earthquake could happen again today?

Unravelling this chain of tectonic events proved far more complicated than I had assumed, or geology textbooks proposed. It required forensics to separate the palimpsest of overlapping tectonic episodes. And the land beneath Britain continues to move today.

★

The idea of the 'Tertiary' (or 'third') geological era was born in Italy around the middle of the eighteenth century, as coined by Giovanni Arduino in 1759. The history of the world had three acts.

'Primary rocks' included the hard crystalline basement of speckled granite or gneisses, gnarled like bacon rashers. 'Secondary rocks' were the thick layers of hardened sediments: the sandstones, the limestones, and shales, often with fossils. Away from the mountains, the layers were mostly flat.

Lying on the Primary and Secondary formations were the formations of the third act—the 'Tertiary'—generally unlithified sands and clays, not yet turned into rock. Most likely, it was thought, these had been deposited during the worldwide flood that had floated Noah's ark.

(Arduino added a fourth 'Quaternary' act to this epic world opera, to include sandbanks in the Po River delta, shifting from season to season.)

To geologize these terms required Greek roots: 'Secondary' became the Palaeozoic Era (old life) before the Mesozoic (middle life), whereas 'Tertiary' was renamed the Cenozoic (new life). In Britain, Cenozoic age sediments, in particular from the second half of the Cenozoic, younger than 30 Ma, were scattered or missing. No one could ever claim Britain was a great place to study the whole of Cenozoic geology, in the way that England and Wales provided a near-perfect cross section through the Palaeozoic and Mesozoic formations. Yet as any crime-fiction sleuth will tell you, just because the evidence is missing, it doesn't mean that 'nothing happened'.

Between the end of the Mesozoic and the beginning of the Cenozoic, 66 million years ago, we now know there was a most dramatic transition, a global catastrophe. With the exception of turtles and crocodiles (animals that lived in water, and whose temperatures fluctuate with their surroundings), no four-limbed animals weighing more than 25 kilograms survived. And that included the dinosaurs—whose lineages had been sustained for 100 million years. Three-quarters of all the species on Earth became extinct. Out went the most spectacular land animals Earth has seen, giant plant-eating dinosaurs, dependent on a massive daily intake of leaves and stems. And once the bodies of the starving plant-eaters had decomposed, the terrifying carnivores, like *Tyrannosaurus rex*, which were at the top of the food chain, also starved and died. The only dinosaurs to survive were the birds.

A range of small mammals then took advantage of this depopulated world, free from larger rivals or predators. Within 10 million years, the ancestors of horses, bats, whales, and primates had evolved to fill all the vacant niches in the global ecosystem.

Meanwhile in Britain, this time also marks the beginning of a completely new chapter. After remaining inert for more than 30 million years, immersed beneath subtropical seas on the north-eastern edge of the central Atlantic Ocean, something was stirring deep in the Earth's mantle, hundreds of kilometres underground. A plume of hotter deep mantle material was rising beneath southern Greenland, expanding its head like a mushroom cloud, and spreading under what was to become northern Ireland and the western British Isles.

Where to begin the history? For most of the territory of Britain and Ireland, the conjoined island (except the fittingly 'nouveau' south-east) first emerged on the scene more or less at the same time as the dinosaurs were wiped out by that asteroid.

★

Offering opportunities for travel and outdoor pursuits, with puzzles to be unravelled and fossils to be unearthed, geology became a pre-occupation, in the first years of the nineteenth century, for a select group of independently wealthy gentlemen. In 1807 some of these enthusiasts established their own London debating and dining club: the Geological Society. It was a year that saw the first passengers transported by railway, the first street lit by gas, the act passed to abolish slavery in the British Empire, and the British bombardment of Copenhagen for taking the wrong side in the Napoleonic Wars.

In the years 1828 and 1829 the dapper young lawyer and aspiring geologist Charles Lyell travelled on the Continent with a *nouveau* rock-enthusiast: Roderick Impey Murchison. Together they visited the celebrated geological landmarks—mountains, volcanoes, fossil outcrops—in France, Italy, and Sicily. Lyell returned convinced that the geological processes around us today—eruptions, earthquakes, floods, and landslides—could explain everything that had happened in the past. There was no need, as argued by German 'vulcanists' and French 'catastrophists', to believe in episodic worldwide catastrophes that raised mountain ranges.

To pursue this idea, that the present was the key to the past, in 1830 and 1832 Lyell published the first two volumes of his *Principles of Geology*. He planned to make a living by writing the 'go-to' textbook on the new science. (Charles Darwin took the first volume with him when he left in the *Beagle* survey ship, at the end of 1831, and had the other volumes forwarded to ports on his circumnavigation.) While preparing the third volume, Lyell proceeded to get married and set out on an extended geological honeymoon through Germany and Switzerland, all the while collecting fossil shells from Cenozoic age sediments. In 1833, in the third volume of his *Principles*, he proposed how to subdivide sediments of the Cenozoic Era into different epochs according to the proportion of mollusc fossil shells that were 'modern' and could still be found on today's beaches. The younger the age of the sediment, the more the fossil mollusc species looked like those alive today.

He divided these proportions into bands and set out to assign names for each band, with suitable, scientific-sounding, Greek roots.

Starting with clays and sands from the start of the Tertiary, in which fewer than 3 per cent of fossil mollusc species were modern, Lyell took the Greek word for dawn and coined Eo-cene, meaning 'Dawn of the New'. This formation was followed by sediments he termed Mio-cene (meaning 'Less of the New'), typically with at least 8 per cent modern mollusc fossils. Then came the 'Old Pliocene' (25–50 per cent modern molluscs) and the 'Young Pliocene' (greater than 95 per cent modern molluscs). (Plio-cene meant 'More of the New'.)[2]

Five years later Lyell changed the name of the Young Pliocene to the 'Pleistocene' (meaning 'Most of the New') to complete the sequence.

Geologists across Britain and Europe soon adopted Lyell's names for classifying the ages of Cenozoic sediments. Then they started to argue how his classification could be improved. In 1854 a German palaeontologist proposed the boundary between the Eocene and the Miocene should be redefined by inserting an additional epoch termed Oligo-cene (meaning 'Few of the New'). And in 1874 a French botanist proposed a 'Paleocene' epoch (meaning the 'Older Dawn of the New') between the Eocene and the Cretaceous.

With these additions, Lyell's intuitive naming convention provides the subdivisions that are still in use today.

Just as we identify eras of British history by way of the Tudors, Elizabethans, Jacobeans, Georgians, Victorians, and Edwardians, the subdivisions of the 'Cenozoic' run: Paleocene, Eocene, Oligocene (together called the Paleogene Period) followed by: Miocene, Pliocene, and Pleistocene (together called the Neogene Period) (Fig. 1). Or, in translations of their Greek roots: 'Old Dawn', 'Dawn', 'Some', 'Few', 'More', 'Most'.

There is a further level of refinement for how geological time is subdivided, down to the 'stage', or 'age', each with its unique name. For a novice this is way too much detail. Yet the classification has its own poetry: the 'Zanclean', the 'Rupelian', the 'Burdigalian', the 'Priabonian'. The list of locations Latinized in these names tells its own story as to where the sediments and volcanics of the Cenozoic were laid down (as interpreted through the myopically Eurocentric lens of nineteenth-century geology).

The first two of the three stages in the Paleocene are Danish, the third refers to an island in the Thames Estuary (Thanet). The first three of the four Eocene stages are from Belgium, north France, and a New Forest town with an erosion problem (Barton on Sea). Meanwhile, the Oligocene is divided between Belgium and central Germany. The six divisions of the Miocene start with two from south-west France, followed by four from Italy, while both the two from the Pliocene are Sicilian. The 'sun-chasing' geological record wanders away from the shores of the grey North Sea and has chosen to hang out by the Mediterranean. We will see what we can do to bring it back.

To subdivide time beyond the broader subdivisions, I will use prefixes, 'Early' and 'Late' (think of first and last thirds), and their equivalents in stratigraphy 'Lower' and 'Upper'. I am going to tell this story chronologically. We will identify time through the Cenozoic using the dial of an 80-million-year clock (Fig. 2).

*

On the northern edge of the Dartmoor plateau, two ancient pathways crossed.

The route from Exeter to Launceston, Cornwall's capital, intersected the 'Mariner's Way' running from the south coast port of Dartmouth to Bideford, an estuary port for the Bristol Channel. Along the Exeter to Launceston path came merchants and clerics,

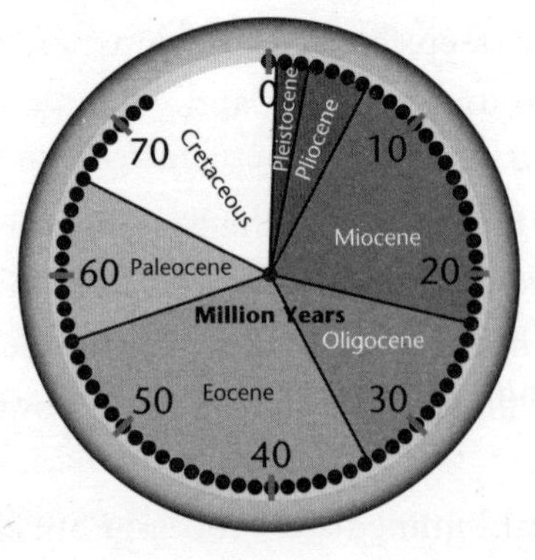

Figure 2. The Geological Time Clock

tradesmen and soldiers on horseback, walking and riding between the ancient county towns of south-west England. Worn into ruts by the wheels of carts, and the horses sliding on the winter mud, the eastern portion of the track was known by its Saxon name 'Harepath'—the path of the Warrior. Along the intersecting Mariner's Way stumbled crewing sailors, hungover from a night in the tavern, as they switched ports between the north Taw and south Dart harbours, according to which coast was favoured for sailing.

West of the crossroads the path to Launceston climbed steeply up the flank of the forbidding granite uplands of the Dartmoor plateau. The village that developed at the crossroads inherited the Saxon name for this ascending trail: Sticklepath—from the 'staecle' or 'steep' path.

In medieval times Sticklepath was Dartmoor's 'northern powerhouse'. The rain-soaked Dartmoor plateau was the first high ground encountered by advancing Atlantic fronts. As the River Taw plunges off the edge of the high Dartmoor granite plateau, it descends into Belstone Cleave, a steep, wooded ravine. In Elizabethan times there were ten water mills in sequence here, powered by the plunging torrent. The valley must have reverberated with the sounds of the great wooden wheels, rumbling and creaking as they turned. The hiss of falling water and the grumble of the water wheels still echo at the eastern end of the ravine, where the Finch Foundry watermill is operated by the National Trust.

The landscape around Sticklepath is moulded by the geology. The torrent has carved a steep, forested canyon where the River Taw passes off the edge of the tough granite and encounters the weaker shales and sandstones.

For hundreds of metres around the 280-million-year-old Dartmoor granite, the near vertical, tightly folded layers of ancient strata were baked by the heat liberated from the vast volumes of granite magma. Superheated water flowed through cracks, depositing veins of copper and tin ores.

But at South Zeal,[3] immediately south of Sticklepath, the rocks up against the granite show no sign of such cooking. These rocks could not have been in this location when the red-hot granite magma arrived. On closer inspection, here the boundary between the granite and the 'country rocks' turns out to be a fault, which has created the linear hollow followed by Sticklepath High Street.

In the early nineteenth century miners opened a copper mine in 350-million-year-old volcanic rocks at Heron's Brook to the south-east of Sticklepath, and became the first to encounter the fault. In chasing the seam to the west they found it came to an abrupt end at the Ramsey Stream. The continuation of the copper seam was later located in the same volcanic layers but dislocated 2.5 kilometres to the north-west.[4] The copper ore lies within a steeply dipping series of volcanic rocks, interleaved with cherts and shales, which continues along the northern fringes of the Dartmoor granite.

It was not until the 1950s that the fault was mapped and named after the village where it was first identified.

Many of the village's buildings are plastered and painted, obscuring the building stone, but the two-storey Sticklepath Stores in the High Street showcases the rocks found on either side of the fault. Dressed granite quoins have been set around the windows and corners of the building, whereas the main ashlar building stone is marmalade-coloured sandstone.

The Sticklepath Fault runs from north-west to south-east. Compare the geology on either side of the fault and, close to the village, the two sides seem to be offset by at least 2.5 kilometres. But the fault is more complex than this single strand. The easterly trending volcanic layers running along the northern side of Dartmoor shift alignment after crossing the mapped Sticklepath Fault, before returning to steer easterly close to Drewsteignton almost 10 kilometres away. In the intervening terrain there are other parallel faults (Fig. 3). Taking the volcanic beds as our yardstick, the total offset could add up to 5 kilometres. But to appreciate the importance of what we find at Sticklepath, first we need to understand some more about these geological faults.

★

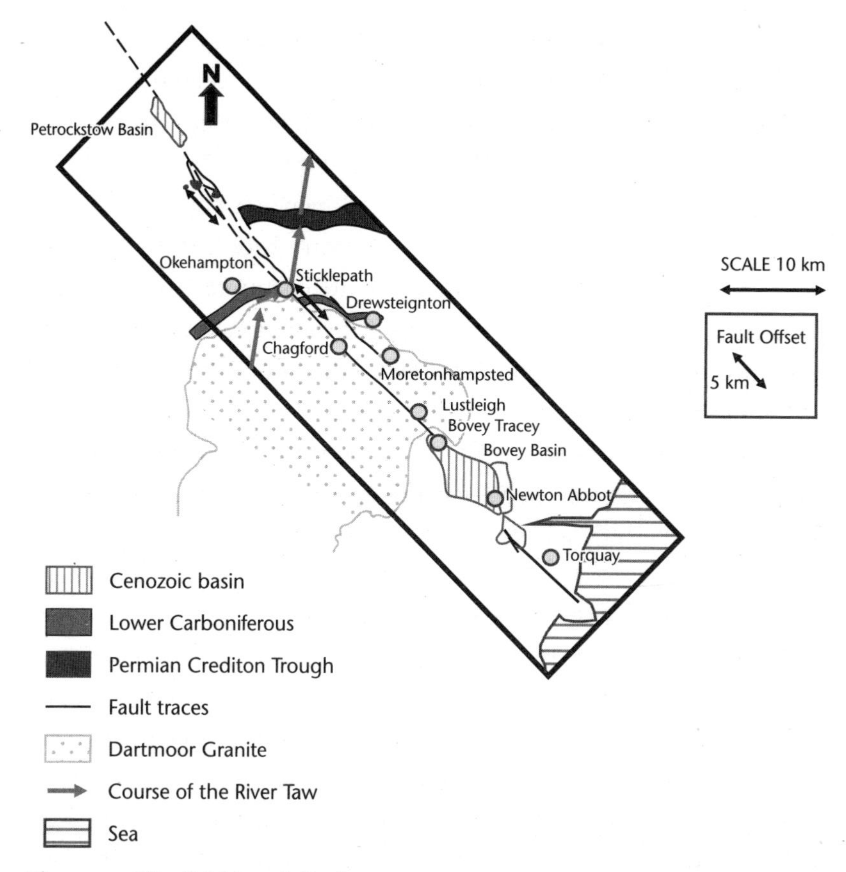

Figure 3. The Sticklepath Fault

A 'fault' is a defect, a misplacement. For the weaver, a 'fault' is where the regular repeating pattern of a woven cloth is interrupted or flawed.

In geology, 'fault' is the name for a sharp planar boundary where the rock layers suddenly come to an end. Typically, on the other side of the fault, the continuation of each layer has been consistently offset.

For miners in their dark, dripping tunnels, the fault was where the precious seam of coal, or vein of tin ore, abruptly stopped. The challenge was then to trace where the seam or lode was displaced on the far side of the fault. Failure could mean the mine would have to close. Yet, if you could find how far along the fault the seam had shifted, you could dig a new tunnel and uncover further riches. A successful miner would learn to 'read' faults, and all the mineral veins and contortions in the layers that accompanied them.

Miners developed a vocabulary to describe what they encountered in their underworld. Like sailors, miners reported depth in fathoms (an old English measure for a pair of outstretched arms). Where the layers in the rock were tilted, they called the slope a 'dip' and measured the angle of dip in degrees, with horizontal 'zero' and vertical '90 degrees'. Lay a horizontal line on the dipping surface and the azimuth in degrees relative to true north was its 'strike'. A sloping fault also had a dip, termed the 'hade'—maybe from the Greek for the underworld, 'Hades'. As they ventured deeper, and the rocks grew warmer, miners suspected their tunnels were venturing into the hellish territory of 'Old Nick', the devil himself, who could play tricks on them, like suddenly hiding the rich seam at the fault. There were even names for the rock on either side of a dipping fault. The 'hanging wall' could form a precarious slanted ceiling to the tunnel, whereas the 'footwall' was a solid rampart.

Deep underground, before cutting a new tunnel to chase the lode, or seam, on the far side of the fault, miners needed clues to help identify where to dig. Perhaps a conspicuous layer had been smeared along the fault? One leading clue for German miners to find how the fault had moved was 'slickensides': polished lines in the material of the fault, where sharper stones had cut grooves in the opposite face, gouged when the fault suddenly moved. Patiently follow the grooves and you should find the continuation of the vein or seam, on the other side of the fault. But these lines would not tell you how far you had to search, how much the fault had moved. And if the fault had shifted differently through its life, the grooves on a single fault could be in more than one direction.

The commonest type of fault is the one where the hanging wall has slipped down the dip of the fault. We call this a 'normal fault'—the standard configuration that you would find at the back of a landslide (Fig. 4). A normal fault forms where the crust is being stretched. Where a pair of normal faults dip towards one another you will find a downfaulted rift.

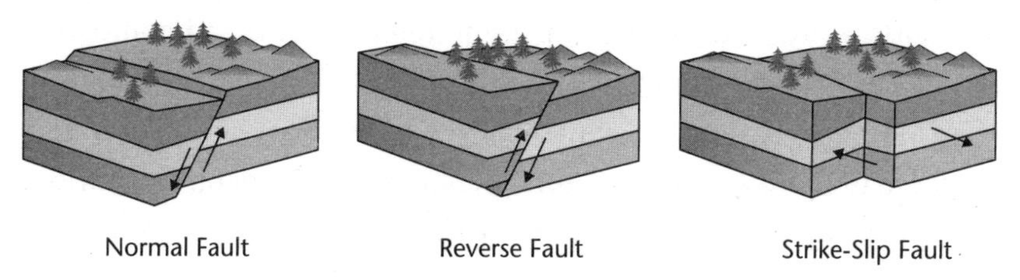

Figure 4. Three classes of faults

The opposite of a normal fault is a reverse fault, formed where the crust is being squeezed or shortened. The hanging wall has been pushed up a dipping plane. The fractures and veins around reverse faults are typically more complex and distorted than normal faults. It takes forceful, horizontal tectonic bulldozing to drive the rock mass up the sloping fault.

In the third category, we have the 'strike-slip fault', where the two sides of a vertical fault have shifted horizontally. The Sticklepath Fault is 'strike-slip'. Such sideways movement may be hard to see unless it offsets something near-vertical or a new strike-slip fault rupture offsets a road or fence.

Big faults are a compound of small faults. The more movement along the fault, the more the constituents of the fault get chewed up and contorted, as the fault zone gets wider and wider. Big faults are like Mongolian desert roads, a kilometre-wide zone of disturbance, within which it may be very hard to say where the latest trail runs.

Big strike-slip faults are also notably shy. Fault displacement brings damage and leaves the surrounding rock a target for erosion. Faults typically become the site of valleys and hollows. It is rare to find a big fault at a natural outcrop. For example, at West Bay, the port for Bridport, Dorset, the bold cliffs to the east expose the prominent horizontal bands of the Upper Lias Bridport sands laid down at the end of the early Jurassic. To the west the cliffs reveal younger middle Jurassic Frome Clay and Forest Marble. Only a significant fault can explain this juxtaposition, yet the fault is nowhere to be seen, having formed the intervening hollow deeply eroded by the River Brit.

★

My wife's family own a four-square miller's house, made principally of cut sarsen stones, with brick around the windows. Built around 1900, the house has been in the family for more than sixty years. The right-hand corner of the house had some slight subsidence on the steep bank down to the stream. Before the foundations were strengthened, the cracks in the interior wall provided an ideal medium on which to contemplate some simple principles about tectonics (Fig. 5).

Close up, we can measure by how much, and in what direction, the two sides of the crack have opened. Where the crack changes orientation, as where it zigzags through brickwork, by repeating the measurements we find the direction and amount of opening are little changed, even though with a new orientation, the crack can appear quite different.

We can apply this lesson to tracing faults and tectonics. At the end of a fault the displacement may have switched onto faults of a different orientation. Along one fault orientation the movement may be sideways; on another, opening and extension. The faults may look quite distinct, even though they achieve the same displacement.

Figure 5. Learning from a crack in a wall

A wide wall-crack does not abruptly come to an end. It may taper, or it may change orientation, but it cannot simply stop within the wall. If a wide crack appears to end, some patient examination may be required to trace where it continues.

Movement tends to get concentrated on one large crack. To open a crack takes work, breaking the fibres and glue in plaster, or the crystalline fabric of concrete, stones, or brickwork. Once a crack has been created it tends to become the focus of any further opening. Reusing the line of an old crack requires much less work than fracturing a completely new crack.

Structural engineers fix 'tell-tales'—little strips of glass—across the largest cracks in the walls of ancient buildings, knowing that this is where future movement will concentrate. The idea is to see if the crack is active and the church tower is threatening to fall. Once a crack has formed, if the load stays the same, the crack will continue to get exercised. That is why 'tell-tales' work.

The principles that work in the wall of a building also apply in the Earth. Faults tend to get reused—again and again.

Cracks on all scales navigate around rigid blocks, as well as follow pre-existing fractures. That is why cracks in the ground, like English country roads, do not run straight.

However, for most of the time movement on a fault is stuck. The two sides cannot slide past each other because the friction on the fault is too great. But the pushing or shearing continues until eventually and suddenly the fault breaks. And it is this sudden breakage and displacement along the rough walls of the fault that generates the vibrations we call an earthquake.

The displacement along a fault rupture typically tails off to either end. A significant crack cannot suddenly terminate. Typically, the shape of displacement along a fault is like a smile—widest in the middle.

Most faults seen in cliffs and quarries are quietly resting, with no immediate prospect that they will lurch into further displacement. Many faults moved at some period in the geological past, when in the vicinity of rifting or plate boundaries. Each fault has relieved accumulated stresses, after which the plate boundary shifts or becomes abandoned. Close to the surface many faults are secondary or superficial, not themselves the source of earthquakes.

However, a large and deep fault is best characterized as sleeping rather than dead. It will remain a zone of weakness. Maybe one day it will awake from its coma into further displacement and earthquake activity.

What we can see in cracks in a house wall can also teach us about plate tectonics. Plates rub up against one another at plate boundaries. The plates could be separating, colliding, or sliding past one another, but their boundaries cannot simply come to an end. We may have locations where a plate boundary changes orientation, and therefore the movement looks different, or may be distributed across more than one fault.

The forces within the plates, pushing and pulling, raising and lowering, are consistent over hundreds and even thousands of kilometres. These forces determine where cracks open and faults break. The forces change over time, but slowly, over millions of years.

These ground rules help us to make sense of past tectonic episodes. The consistency and slow rates of change mean that individual tectonic clues can be extrapolated over large distances and long time periods. There is a strong logic to tectonics. In the course of Britain's journey through the last 66 million years, we will encounter a handful of tectonic configurations that have become widely duplicated at the same period, like finding an identical stitch repeated in a work of embroidery.

★

Naming the longest faults can present quite a challenge. Give the fault a local name, after the village where it was first mapped, and you may find it has wandered over moor and mountain and become someone else's local fault, with a different name. (Rivers used to share this problem, having different local names along their course—or simply being labelled 'river', as in the River Avon, meaning River River.)

For the long-distance strike-slip faults of Devon and Somerset, rather than fight for priority the name gets compounded. The 'Sticklepath–Lustleigh' Fault is the full name for the south-east extension of the Sticklepath Fault. The triple-barrelled 'Watchet–Cothelstone–Hatch Fault' runs parallel to the Sticklepath Fault 80 kilometres to the east, through the county of Somerset, emerging in a baroque display of fractures and gypsum veins in cliffs to the west of the charming Bristol Channel port of Watchet. Like the multiple surnames of the most distilled British aristocrats, beyond three names this would start to become tiresome.

Between the towns of Sticklepath and Lustleigh, along the edge of the Dartmoor granite, strands of the Sticklepath Fault have created tectonic landscapes that would not look out of place in coastal northern California, with its Pacific v. North America strike-slip plate boundary. The River Teign drains eastern Dartmoor and then wanders for 10 kilometres through an open rolling landscape with a broad flood plain, before passing through a 150-metre-high steep fault wall that has risen across its path. The river has had to carve a gorge for another 6 miles through the massif. Early in the twentieth century a wealthy store-chain owner, Julius Drewe, had 'Britain's last castle', 'Drogo', constructed on a rocky bastion at the lip of the gorge, taking advantage of the terrain and the views sculpted by tectonics along the Sticklepath–Lustleigh Fault.

Further south, the Bovey River drains more rolling open landscape before plunging south-east through another forested gorge: Lustleigh Cleave. After 5 kilometres in the ravine the valley opens out once again. The fault has juxtaposed hard granite against weaker strata. The raised ridges and lowered depressions create surprising and beautiful landscapes.

As we chase the Sticklepath–Lustleigh Fault to the south-east, before it slips into the English Channel it passes straight through Torquay, proud capital of the Devon Riviera and home to the fictional Fawlty Towers. Residents might be disconcerted to learn that a prominent strand of this prime fault passes into the town at Chapel Hill and runs down the Rock Walk in Royal Terrace Gardens, behind the former Palm Court Hotel.[5] Beyond Torquay the fault continues to the south-east towards Brittany, shifting its alignment mid-Channel.

Eighty kilometres away, at its north-west landward end, the Sticklepath Fault passes demurely into the Bristol Channel, in the vicinity of Bideford. Curiously, there is a dell named Babbacombe close to both ends of the onshore fault. (Who was Babba, and how did she get to name these little far-apart twisted valleys?)

Where the trace of the Sticklepath Fault dives underwater it continues far offshore to the north-west, past Lundy island. A branching fault intersects the western tip of Pembrokeshire, but the master fault makes it almost to Ireland's south-east corner at Carnsore Point. The fault can be traced farther offshore than its course on land. Its total mapped length is around 300 kilometres.

So when did the two sides of the Sticklepath Fault slide horizontally past one another? To the north-west of Sticklepath village, the famous fault intercepts the Crediton Trough, an E–W rift valley filled with lithified desert sand dunes, more than 250 million years old. Several strands of the Sticklepath Fault cut the trough. The south-west side (the 'Hatherleigh Outlier') has shifted up to 5 kilometres to the north-west. So the displacement on the fault is younger than 250 million years. As it happens, we can constrain this timing much more precisely.

*

Like road etiquette and gloves, strike-slip faults with horizontal displacement are either right- or left-handed. Stand facing the fault and ask the question: has the other side of this fault moved to the right or to the left? If it has moved to the right this is a 'dextral' (right-handed) fault, to the left and it is 'sinistral' or left-handed (Fig. 6).

The strike-slip San Andreas Fault in California is dextral. Stand in San Francisco facing the fault that runs offshore to the west of you, and the far side—the Pacific plate—is moving to the right, to the north. Stand in Santa Barbara, west of Los Angeles, and face the San

Left-handed (sinistral) and right-handed (dextral) faults

Left–lateral slip

Far side of fault moves to the left: 'sinistral'

Right–lateral slip

Far side of fault moves to the right: 'dextral'

Figure 6. Sinistral and dextral strike-slip faults

Andreas Fault that runs to the north-east of you, through the San Bernardino Mountains. Santa Barbara is on the Pacific plate, whereas the inland city of Bakersfield, on the American plate beyond the fault, is moving to the right—to the south-east.

The 5-kilometre offset of the Crediton Trough proves that the Sticklepath Fault has moved dextral strike-slip. Farther south-east along the same fault the 400-million-year-old Devonian Torquay limestone has also shifted 5 kilometres dextral from its matching East Ogwell limestone.

However, the Dead Sea Fault, which passes due south through Syria, Lebanon, northern Israel, along the border between the West Bank and Jordan, and then along the Jordan–Israel border down to the Gulf of Aqaba, is 'sinistral'.

Faults are not simply flat, like a football pitch. They grow out of pre-existing faults and cracks, so they are irregular and uneven. A sudden displacement can jump from one line of fracture to another. And sometimes the whole line of the fault shifts to a parallel fault, offset by hundreds of metres or even several kilometres. When this happens the simple sideways movement along a strike-slip fault starts to get complicated.

Geologists are never happier than when drawing on the bare earth with a freshly cut stick or walking pole. First mark a pair of parallel lines to represent the traces of two faults at the surface. The movement on these faults is horizontal. The displacement starts on one of these lines and then midway it jumps to the other parallel fault.

The jump in the line of the fault is called a 'jog'. Stand on the trace of the fault and look towards the jog. The jog could be either to the left or to the right (Fig. 7).

If the jog and the fault movement are contrary to one another, for example, a left jog on a dextral fault, then the jog becomes a crunch zone, or a push-up—the two sides colliding with one another. There is nowhere to go but up.

If the jog is to the right and the fault is dextral, then the jog opens into a gap or hole. The same happens with a sinistral fault and a left-handed jog.

Initially extension may be carried by a single normal fault that straddles the gap. The hole typically develops a coffin shape, with the long axis following the gap between the overlapping faults, called a 'pull-apart basin'—or 'pull-apart' for short (Fig. 8). As movement jumps across the jog, again and again, in repeated fault movements, the hole continues to be pulled apart and the ground surface sinks.

Typically the hole becomes a lake. The most famous pull-apart basin is the Dead Sea between Jordan and Israel, which occupies a

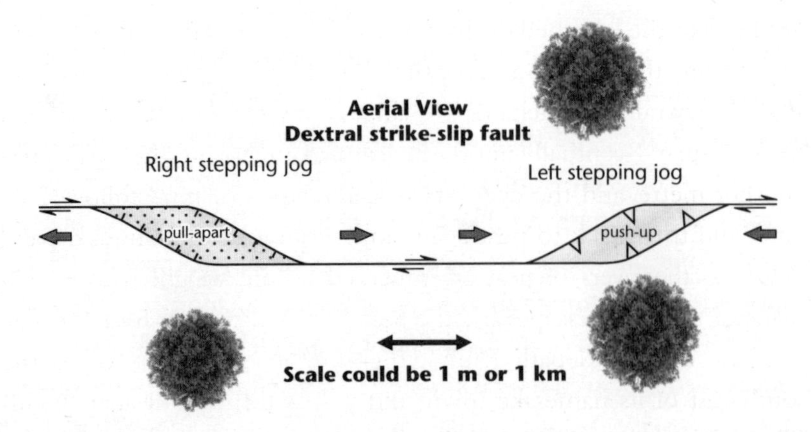

Figure 7. Right- and left-handed jogs on strike-slip faults

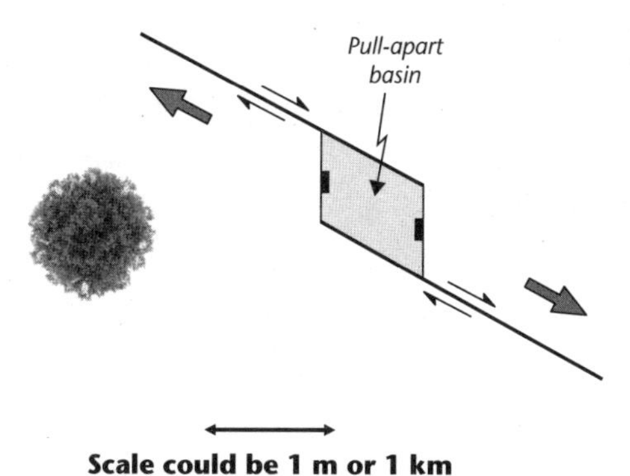

Formation of a pull-apart basin. Aerial view, dextral strike-slip fault, right-hand jog. Grey marks the hole.

Figure 8. The formation of a pull-apart basin

cigar-shaped north–south basin, 130 kilometres long by 10 kilometres wide. The controlling N–S 'sinistral' strike-slip faults are offset to the left. Low rainfall and strong evaporation have lowered the lake surface to 400 metres below sea level. The Dead Sea is a chemical soup with a recipe made up of one-third dissolved salts, more than eight times the salinity of the ocean.

More typically, the rivers flowing into a pull-apart fill a freshwater lake and transport rock fragments, sands, and clays eroded from the surrounding landscape. A delta of braided channels forms where the river water stills as it passes into the lake. Trees grow among the marshes. When, eventually, the strike-slip faults to either side slip in an earthquake, the lake-bed drops, the forests drown, and the dead trunks topple into the swamp and, in the absence of oxygen, turn to peat. After heavy rains, the delta river changes course and the sand and mud banks return. Eventually another earthquake happens, the basin sinks another metre, and the cycle starts again: layers of mud followed by sand, and then back to peat again, in a sequence of earthquakes and deltas. As the layers of peat are squeezed by the weight of overlying material, the water escapes and the peat turns to lignite, halfway to coal.

Chase the Sticklepath Fault either to the north-west or to the south-east of its namesake town, and you will find pull-apart basins filled with sediment (Fig. 3). To the north-west at Petrockstow this pull-apart basin measures 7 kilometres long by 1.5 kilometres wide, and has been core-drilled to its base more than 700 metres down. To the south-east of Sticklepath lies the larger Bovey Basin which today measures 11 kilometres by 8 kilometres, but which has only been cored to 40 per cent of its estimated 1,200-metre depth.

In the 1980s geologists experimented with scaled-down sandbox models of pull-aparts to find the link between the sideways movement and the sinking. They found the floor of the pull-apart sank an average of 15 per cent of the horizontal movement along the bounding faults.

Using this figure of 15 per cent, the 700-metre depth of the sediments at Petrockstow would convert into 4.5 kilometres of displacement on the Sticklepath Fault. Applying the same formula to the Bovey Basin would imply as much as 8 kilometres of horizontal displacement, inconsistent with other observations. This suggests in this broader basin there was a higher-than-average ratio between pull-apart depth and horizontal displacement.[6]

When these basins were sinking, the climate was subtropical. The hot and humid conditions decomposed the rock into a pure white 'china-clay'. The clays became washed by streams into the Bovey and Petrockstow pull-aparts, where today they are quarried for making a range of fine pottery and porcelain.

Among the lignite seams in the basins are spores of the trees and plants and the skeletons of the animals alive when these pull-aparts were lakes. From these fossils we can date the sediments. Samples from the base of the Petrockstow Basin are around 52 million years old (the early Eocene), whereas the strata near the surface in the Bovey Basin are close to 25 million years old (towards the end of the Oligocene). The sinking of each pull-apart was directly linked to movement along the bounding strike-slip faults. So now we know the timespan when the Sticklepath Fault was most active: long after the death of the dinosaurs. In the 'shallows' of deep time.

*

The 1,300-kilometre-long San Andreas Fault cuts through California, linking Los Angeles and San Francisco. The magnitude 7.8 earthquake that occurred here in 1906 was caused by a sudden horizontal dextral displacement along 480 kilometres of the northern San Andreas Fault. In coastal northern California the displacement offset roads, fences, and barns by as much as 6 metres.

The nearest big, active, on-land strike-slip fault to Britain today is in Turkey. It is called the North Anatolian Fault, because it runs east–west all the way through northern Turkey for more than 1,100 kilometres. And along the way there are many places where the fault shifts its alignment and has created pull-aparts, occupied by a lake or swamp underlain by a thick pile of mud and sand, transported by rivers from the surrounding uplands. Big earthquakes, involving fault movements of a few metres, occur every few centuries along one or another section of the North Anatolian Fault, refreshing the pull-aparts with a renewed pulse of subsidence. In 1912, 150 kilometres of the western end of the North Anatolian Fault ruptured, with the displacement passing straight through two pull-apart basins.[7] And, it should be noted, the North Anatolian Fault is a plate boundary between the Turkish and Eurasian plates. Looking farther afield, we can find other long strike-slip faults with deep pull-apart basins, which have accumulated kilometres of movement: Alaska, Sumatra, Western China, Guatemala. In all these locations the faults are components of a plate boundary.

The Sticklepath Fault should be up there in the pantheon of notable strike-slip faults. As we shall learn, it is linked with other faults which extend its length to more than 600 kilometres. Like other great strike-slip faults, has it also been part of a plate boundary?

Imagine finding an overgrown section of eight-lane motorway about which nothing was formerly known. Or uncovering tens of kilometres of high-speed rail-track buried under the leaf mould in a forest. What was a component of a plate boundary doing, active 'the day before yesterday' (in geological terms), in bucolic rural Devon?

Our best estimate for the total dextral displacement on the Sticklepath Fault is 5 kilometres. A typical earthquake on such a fault might be magnitude 7, caused by a fault rupture 50 kilometres long with 2–3 metres of displacement. This was the size of earthquake that devastated the city of Port au Prince, the capital of Haiti, in January 2010. It is close to the dimensions of the fault rupture that occurred to the west of Christchurch, New Zealand, in September 2010. To achieve the total displacement on the on-land Sticklepath Fault required an estimated 2,000 magnitude 7 fault ruptures and accompanying earthquakes. Consider what their impact would have been if this land had been populated as it is today: mass destruction, thousands of fatalities.

How should the inhabitants of the frontline towns and villages of mid-Devon, villages like Chagford, Moretonhampstead, and Lustleigh, but also Torquay and Bideford, view this magnificent fault? Is it really dead? Perhaps it is only sleeping? Volcanoes are classed as 'active', 'extinct', and 'dormant'. Is this a dormant fault? Could one safely give it a poke? Are there still jolts to be expected along it? How can we be sure it won't suddenly rear up and bite us? When was the last time it caused a big earthquake?

Along the San Andreas Fault in California one of the most characteristic features of its recent activity is the displacement of stream gullies, all shifted consistently to the right. Sometimes a gully has shifted 6 metres in a single earthquake, sometimes it is 100 metres accumulated from many repeated displacements.

On Dartmoor, the valley of the River Taw trends consistently a few degrees to the east of north. Downstream of Sticklepath the river maintains the same direction in its path towards the north coast of Devon: a few degrees east of north (Fig. 3). These two alignments are offset from one another by almost 2 kilometres. However, measured along the line of the fault the offset is more like 3 kilometres. It is as if the line of the river precedes much of the shift on the fault.

Even more notably, after the torrent plunges down Belstone Cleave, the river makes a sharp right turn for a few hundred metres at Sticklepath village. Could this in turn reflect the last few episodes of fault displacement?

★

If I were writing a book about the past 60 million years of tectonics along the east coast of North America, the work would be short. Not much happened.

On the European side of the Atlantic the tectonic history is as complicated as the political history. A lot has happened. The tectonic scene has shifted repeatedly. To unravel this story we need forensics and deduction. We need to pay careful attention to ('radiometric') dates, but also understand how such dates can be prone to errors. Above all we are guided by two key principles: that the overall (synoptic) stresses are consistent across a region, and that a big displacement on a fault, or crack, cannot simply come to an end.

When television 'does' geology, the camera soon gets bored of an immobile rock, and switches attention to an inquisitive marmot or gets the presenter to be lowered down a rope into a cavern, crevasse, or crater. In this story some animals will make an appearance, along with horizon-stretching forests. There will be detours to appreciate artefacts ceremonial and functional. We will also be in the company of some enthusiastic scientists, receiving several guest appearances from Charles Darwin.

We will start this story somewhere which does not need to be gussied up with marmots or rope tricks, on a visit to the most spectacular example of Cenozoic tectonics in Britain. Conveniently, it can be accessed by chairlift.

2

The big tilt

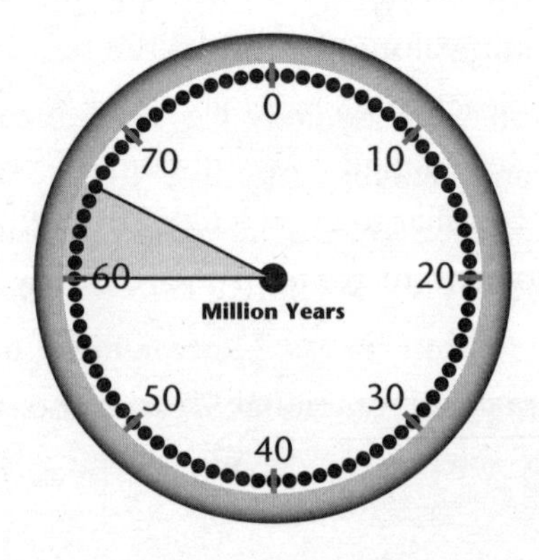

Like most big faults, the fault at Sticklepath is poorly exposed and unspectacular. You can't stand in front of an outcrop and see the displacement. Only perhaps in Japan would they have cut a great trench, dated the last movement, and built a museum to house the hole in the ground, as at the fresh fault scarp (the first ever photographed) created by the devastating 1891 Mino-Owari earthquake in central Japan. I discovered the new museum in 1995 after arriving on a child-sized bicycle, the largest available to rent from the nearest railway station. While nominally attending a conference in Kyoto, I had bought a Japan rail-pass to spend a few days visiting the latest fault scarps across Honshu. But there is another example of geologically young tectonics in Britain, contemporary with activity on the Sticklepath Fault. Here is a fault that can really be seen, so much so that it is a tourist attraction.

At the extreme western end of the lozenge-shaped Isle of Wight, the island narrows to a hilltop peninsula surrounded by tall and precipitous cliffs. For 100 miles along the coast of southern England, wherever the rock is exposed in cliffs you will find uninterrupted layers of clays and limestone, lying within one or two degrees of horizontal. Yet at the western tip of the Isle of Wight the layers are all close to vertical (Fig. 9). The world has been rotated onto its side.

In the western extension of that peninsula, in a series of wild gothic rock stacks continuing along the line of the headland, the towers and cliffs are made of bleached white chalk. 'The Needles' are part of this 'world tipped on its side' geology, upstanding precisely because the layers are vertical. Familiar to every sailor, these white spires are manifestations of exuberant tectonics. A particularly durable limestone

Figure 9. The near-vertical Paleogene strata at Alum Bay, Isle of Wight

band has remained hoisted into the air, even after all the surrounding chalk has been peeled by the battering waves.

To the north of the peninsula lie the cliffs of Alum Bay in which the 'world tipped on its side' offers a most unusual way to experience geological time. What were once horizontal layers of sediments are displayed vertically like a shelf of books.

You can move up and down the section, and move up and down the millions of years of the Paleogene, simply by walking along the beach of Alum Bay at the foot of these cliffs—north to younger times, south back in time all the way to the Cretaceous chalk.

The layers laid down through the first 30 million years of the Cenozoic are surprisingly diverse and colourful, and include the famous mottled bloody reds of the Reading Beds. One layer of pure white sand was formerly mined and shipped to London, Bristol, and Worcester for the manufacture of glass and porcelain.

John Betjeman encouraged us to:

> Walk along the shingle to the arm of Chalk, and then look back at these capes. One is brilliant gold; the next white; the next purple; the next grey; the next black with streaks of green. In great broad bands these strips of colour run down the cliffs, turning the sky pale in the richness of their colour. My Victorian guide book says of Alum Bay 'one side of it a wall of glowing chalk, and the other a barrier of rainbows'.[1]

The beach access to this extraordinary section through prehistoric time launched a Victorian tourist craze. The earnest visitor would take a glass phial and carefully fill it with layer after layer of the different coloured sands and clays, exposed in the cliffs. In his 1844 *Handbook to the Isle of Wight* Thomas Brettell claimed Alum Bay was unique in having twenty-one recognized shades of coloured sands. The geologically minded could follow the actual sequence, sampling from south to north along the beach, placing the oldest chalk at the base of the glass and the youngest Oligocene at the top. Then they would own a microcosm of the first half of the Cenozoic, from the death of the dinosaurs to the rise of the primates.

Photos from the beginning of the twentieth century show hundreds of people on the beach, ritually sampling the vertical sediment layers. In 1860 a bottled set of the Alum Bay layers was presented to Queen Victoria at her residence on the Isle of Wight. There is nothing like this anywhere else in northern Europe. On a sunny afternoon, with your back to the sea, you could imagine you were in Zion Canyon, Utah.

(Today, 'Health and Safety' has intervened in this innocent pastime, and for fear of rockfalls, sand collectors are required to fill their glass phial at the tourist shop. Some of the geological romance has gone.)

This world on its side at the Needles and Alum Bay[2] is another manifestation of the big tectonics also seen along the Sticklepath Fault. The forces that tipped over the rocks in a giant fold at the Isle of Wight came from underneath, along a great reverse fault, with the southern side raised up at least 550 metres.[3]

★

In July 1635 Robert Hooke was born in a farmhouse at the foot of Hook (now renamed Hooke) Hill in the village of Freshwater, at the western end of the Isle of Wight.[4] To the south of the village lay the great tectonic fold, the finest example of recent tectonics in southern England. The cliffs of Alum Bay were a short walk to the west.

Something of the geology of this location got into his character and obsessions.

Sickly as a child, Robert showed a prodigious appetite for learning and experimenting. We know he roamed the local clifftops and beaches, collecting fossils, tracing the strata, and observing the erosive force of the sea. Many years later he told Robert Boyle about clifftop observers gesticulating to the boats below, to point out where the fishermen should throw their nets to catch shoals of pilchard. From the clifftop you can see deep into the translucent water, while on the boat there is nothing but glare. Hooke told Boyle that from a clifftop he could see 'how far the rock ran out into the sea'. A western spike of the Needles chalk rock stacks, 'Lot's Wife', protruded 40 metres above the water (until felled by a storm in 1770).

After attending Oxford University, in July 1664 Hooke became the new Royal Society's 'Curator of Experiments' with a precocious list of ambitions: to work on the development of 'theories of Motion, Light, Gravity, Magneticks, Gunpowder, of the Heavens; an Inquiry into the figures of Bodys; Improving Shipping, Watches, Opticks, Engines for Trade, Engines for Carriages'. (Hooke was later known as 'England's Leonardo'.)

In 1668 Hooke gave the first in a series of lectures on geology. He explained that fossils were the remains of formerly living organisms. Fossiliferous layers had been raised out of the sea through the action of earthquakes.[5] (In his time 'earthquake' could include eruptions and landslides, as well as land movements accompanying the quaking earth.)

Of all the instances of fossil shells found inland across England:

> in the Hills, in the Plains, in the bottoms of Mines and in the middle of Mountains and Quarries of Stone...Now tis not probable that other Men's Hands, or the general Deluge which lasted but a little while, should bring them there; nor can I imagine any more likely and sufficient way than an Earthquake, which might heretofore raise all these islands of Great Britain and Ireland out of the Sea.[6]

Robert Hooke recounted the observations he had made close to his home at Freshwater of a horizontal layer of mussel shells in the cliffs, 50 feet above the beach level. Only earthquakes and eruptions, he claimed, had the power to transform sea to land.

The cliffs at Freshwater gave Hooke an insight into the lengths of geological time. Great changes had driven some species of fossil sea animals to extinction, leaving no living examples. Meanwhile, new varieties of creatures had emerged in response to the shifts in the land and the climate. Here was the germ of an idea that would eventually lead to a theory of evolution. Species became adapted to their climate and food resources and might go extinct when these shifted.

At the time he presented his lectures, Hooke's ideas were completely original: that earthquakes had both lithified fossils and raised them out of the sea. Yet his lectures on earthquakes and geology were not published in his lifetime, for fear of being considered heretical in their challenge to the account of Creation and the Flood in the Book of Genesis. The lectures were rediscovered late in the eighteenth century by Rudolph Erich Raspe, the picaresque author of the Baron Munchausen tales. It was not until 1831, in the first volume of his *Principles of Geology*, that Charles Lyell acknowledged that Robert Hooke's treatise on 'Earthquakes' was 'the most philosophical production of that age'. At least a century ahead of his time, Hooke had recognized that prehistoric earthquakes had raised and contorted the geology of Britain.

★

William first noticed the little petrified fossil seashells in the newly ploughed fields. The local name for them was idlibs. He was also fascinated by the poundstones, used by the milkmaids for weighing milk and cheese, and realized they were formerly five-sided sea urchins, each about 10 centimetres across.

Little had changed for hundreds of years when, on 23 March 1769, William Smith was born in the village of Churchill, Oxfordshire. The industrial revolution was under way, but had yet to ruffle the labour and routine of a farming village. William was a very observant boy, with an extraordinary visual memory that enabled him, even in old age, to remember all the details of his boyhood. His artisan father, a 'very ingenious mechanic', died when he was only eight years old, after which he went to live with his farmer uncle. When he was twelve or thirteen, he travelled to London and saw a newly dug cutting on the road over the hills to Henley, in which the rock exposed was the unmistakable pure white chalk.

The teenager so impressed a visiting surveyor, Mr Webb, that he invited the boy to become his apprentice. William quickly learnt the instruments of his new profession and set out to find every opportunity to expand his horizons.

Canals were all the rage, and there was a boom in survey work. In 1791 Webb sent the twenty-two-year-old to an estate at Stowey in

Somersetshire, whose owner had requested a surveyor to evaluate their underground coal reserves. William made the 120-kilometre journey on foot. At his destination he took a spade and dug into the earth, and was excited to find, beneath the soil, red marls, which he instantly recognized were identical to the marls he had seen below the soil in Worcestershire.

As he climbed down the long, rickety ladders into yet another coal pit, barely lit by the oil lamps, he saw that the arrangement of the seams of sands, mudstones, and coal, even down to the thickness of the layers, stayed very constant even while the layers were strongly tilted and faulted. Miners knew their way through this sequence whenever they encountered it, as though they were the numbered pages of the same section of a book, naming the strata 'Dungy Drift' or 'Temple Cloud', recognizing when the 'Slyving' overlay the 'Firestone'. The tilted and faulted layers in which the coal was found were truncated by an overlying, near-horizontal sequence of limestones and shales (Fig. 10).

In 1793 William Smith was appointed lead surveyor for a new canal to transport the coal from the Somerset mines onto the Kennet and

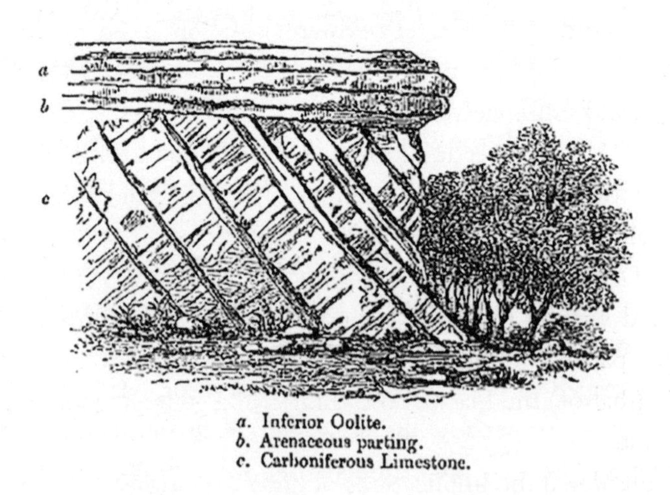

The great unconformity figured by De la Beche

Figure 10. Jurassic strata above unconformity on tilted Carboniferous strata, Somerset

Avon Canal, which linked London with Bristol. In that same year, aged twenty-four, he was found to have written in a notebook that a 'speculation which had come into his mind regarding a general law affecting the strata of the district, was submitted to proof and confirmed'.[7]

He had discovered that the strata lying above the distorted coal seams were consistently inclined at one or two degrees down towards the east, as though the layers had, 'on a large scale, the ordinary appearance of superposed slices of bread and butter'. In summer and autumn 1793 he had conceived this to be 'a general law', constructing a test of his hypothesis by accurately predicting the projected height of three strata in two parallel valleys, intended as routes for the canal.

In 1794 the Somerset canals received parliamentary approval to proceed. Wishing to harness the latest technology for both their canals and their coal mines, the scheme's investors sent two engineers to accompany 'Mr Smith' on a 1,500-kilometre tour of other navigations in England and Wales. He particularly appreciated the times when the carriage had to slow to a crawl up a steep hill, as he could alight and inspect the geology exposed by the roadside. His vivid memory preserved a record of every facet of the rocks he observed. He identified in the Hambleton Hills of north Yorkshire the familiar features of the Cot(te)swold Hills, viewed from the Vale of Gloucester. Climbing to the top of York Minster and looking out 15 miles to the east, from the contours of the hills he recognized that the Wolds were capped with chalk.

On 2 December 1796, while staying at the Swan Inn at Dunkerton, he started writing his observations into what he hoped would be a scientific paper. 'I found the inclination of every bed to be nearly the same as (that of) the Lias... everything had a general tendency to the south-east.'

He considered the implications of his observations:

> ...is it not reasonable to suppose that the same strata may be found as regular on one side of a sea or ocean as on opposite sides of a deep

> valley upon land, and if so, and the continuation of the strata is general, what is their general direction of drift? Is it straight lines from pole to pole, or in curved lines surrounding the globe regularly inclined to the east?

He explored some general cause for the assumed regularity but admitted 'all theories are best built on practical rules'.

The scientific paper was never completed.[8]

His attention to the tilted layers had taught him to identify the unique fossils that each formation contained. Invited to approach a jumbled pile of fossils in a cabinet of curiosities, he would reorder them into their precise stratigraphic sequence. As he put aside his theory of the 'Big Tilt' to concentrate on his role as full-time surveyor on the Somerset Coal Canals, he also started and failed to finish a scientific paper showing how the layers of a particular age can be correlated according to the fossils they contain.

It was this second intuitive theory that would bring 'Strata Smith' fame for his stratigraphic map-making. For the next thirty years, he went on to work at the practical intersection of civil engineering and geology. In a typical year he travelled more than 15,000 kilometres, much of it on foot so as not to miss any geological observations, all the while collecting additional detail to add to his geological map of England and Wales, first published in 1817.

Meanwhile he completely abandoned his investigations of why the younger strata of England sustained a consistent tilt, and how this phenomenon could somehow be explained by the rotation of the Earth. (In fact, Smith was on to some physics that attracts keen research today—that in the Earth's rotation, the outermost rigid shell, the 'lithosphere', is not quite keeping up with the underlying hot and mobile mantle: hence why in the Pacific the subduction zones are hard up against the continents to the east but have drifted out into the ocean along the island arcs to the west.)

Yet England's big tilt is real. There are places where the slight dip down to the east can actually be seen. Stand out on the Cobb harbour wall at Lyme Regis, look to the east and the Jurassic strata tilt away

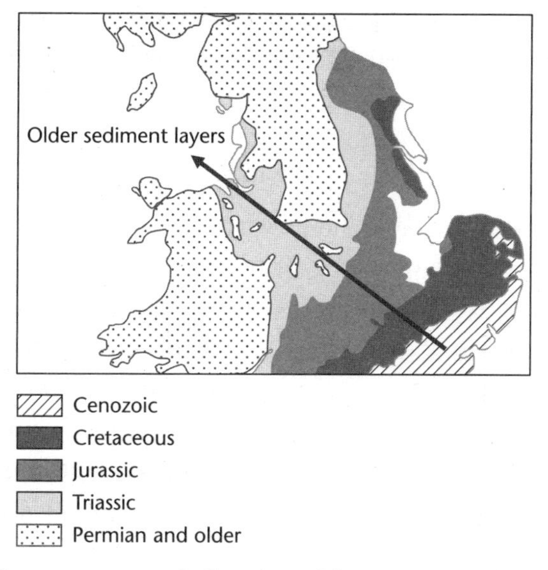

Figure 11. Younger strata encircling the uplift centre in north-west England

from you. Along the lower River Severn, in places the river has carved long, low cliffs which can be viewed from the opposite bank. At the Garden Cliff, at Westbury-on-Severn, the prominent tea-green and rust-red 'Rhaetic' layers (continuing up to the beginning of the Jurassic) dip gently and steadily down to the east.

We now know the easterly slanting stratigraphic layers (like 'superposed slices of bread and butter') that define the geology of much of England owe their origin to something that happened to the western side of the British Isles at the very beginning of the Cenozoic era. The whole of the British Isles was slowly and permanently raised in a vast elongated dome, and England and Wales comprised only the eastern half of this uplift (Fig. 11).

★

When in 55 BCE Julius Caesar clambered into his boat to make the crossing from Boulogne, he was sailing towards a thick white line on the horizon. Britain was a land guarded at its nearest approach to the continent by colossal white walls. Even in ancient Greece, 500 years before the modern era, this island was known as 'Albion', we can only assume from the word-root for 'white', which also gives us 'albino' and 'albatross'. That this island should be known as the white land makes perfect sense if you are taking the shortest journey across the English Channel.

When the Romans returned a century later, most of the land they first occupied was underlain by chalk. Extending to the west through Salisbury, north to Lincoln, and inland to the edge of the Chilterns, chalk ridges marked the skyline. The Icknield Way, known as the 'oldest road in Britain', ran from the coast of Norfolk to the coast of Dorset, along the foot of the chalk scarp. Beyond the chalk the clay soils made the winter roads impassably muddy. (The Glastonbury music festival is sited on the wrong geology: glutinous, mud-churning Jurassic clay.)

For much of south-east and east England, the ploughed fields are thick with knobbly flints, valleys are dry, and the forest floor is dark beneath the thick beech canopy. The chalk rock is never far beneath the surface. Chalk contained plentiful flint: to be knapped for blades and arrowheads, and clashed against iron for sparks of fire.

In the centuries before the Roman invasion, Iron Age men and women cut elaborate dyke and ditch fortifications around the summits of the most prominent chalk hills, leaving white scars ringing the hill. Sometimes they carved two or three lines of such fortifications, in what seems to have been as much landscape art as lines of military defence.

When the Romans finally invaded in 43 CE, England and Wales were still divided into tribal regions. The Romans played the tribes off against one another and within two years all had come under Roman dominion. Within a generation, as the tribespeople were progressively subsumed by Roman laws and Roman armies, even the tribal names had gone. Only for the names of two tribes across central and southern Wales, the Ordovices and the Silures, there would be a strange destiny, almost 1,800 years after their people had been absorbed into the Roman empire.

The chalk rock of the cliffs of Dover, the chalk of the Lincolnshire Wolds and the cliffs of Flamborough Head, the 'white limestone' of the coast of Antrim, and the chalk exposed in the ring forts of Dorset were deposited over a period of almost 30 million years, from 100 to 70 million years ago.

Sea levels during the time of the chalk seas were higher than at any time during the last 600 million years. The supercontinents, such as Gondwana to the south, were breaking up, with new spreading ridges and shallow oceans, displacing the ocean water to drown the low-lying continents. Carbon dioxide levels were four times higher than today and there were no ice sheets at the poles.[9]

Chalk is the equivalent of empty pages, blank video tape, white noise: a continuous rain of microscopic coccolith skeletons falling from the clear blue water. The absence of sand and mud in the chalk means that there was no significant nearby land, no cliffs or streams from which a pulse of sediment would come following a storm. The sea overlying the chalk ooze was clear and blue, like today's reef-fringed lagoons of the Bahamas, far from any eroding landmass.

Even while the largest land animals ever to inhabit the planet wandered through the forests and swamps of North America and China, north-west Europe remained resolutely underwater.

Hundreds of metres of chalk survive across eastern and south-eastern England. Did a similar layer once cover the whole of the British Isles?

Chalk is made of calcium carbonate, calcite fragments, which over geological time dissolve in water to leave nothing behind. How would we know a thick chalk covering had ever been there?

As we travel to the west and north-west, older and older layers are exposed at the surface. And yet in Antrim, Northern Ireland, up to 150 metres' thickness of the Ulster White Limestone Formation (the local name for chalk) has survived. The area was stretched and sank as it was covered with layer after layer of basalt lavas early in the Cenozoic. The lavas shielded the chalk from further erosion and marinated the weak rock with warm water, causing the chalk to recrystallize into harder limestone.[10]

In southern Ireland the bedrock is much older. Any chalk layer would have been situated high above the present landscape. Against all the odds, in the Gweestin Valley, near Killarney, County Kerry, a 2-million-ton block of chalk has plunged an estimated 900 metres down some vast, long-lost limestone cavern to survive as the 'Ballydeenlea outlier'.[11] This boulder is powerful evidence that the whole of Ireland was once covered with chalk as well as riddled with underground rivers.

In south-west Scotland, once again any chalk layer would have been hundreds of metres above the present land surface. Yet somehow here too a monster chalk boulder has survived, this time because it fell down a crater, that of the central volcano on the island of Arran.[12] The easily eroded chalk has today become 'Pigeon's Cave'.

The only part of the British Isles where we can show there was not a thick chalk covering is in north-western Scotland. At Gribun on the Ardnamurchan peninsula, throughout the time that thick chalk was being deposited farther south no more than 20 metres of sediment was laid down, including 3 metres of chalk with the remainder lithified beach sands of the finest purity—99.8 per cent silica. All the minerals which typically bring colour to a beach had been dissolved by the warm Cretaceous sea. Since 1940, at Loch Aline near Morvern, 50 kilometres of tunnels have been cut to mine this silica sandstone for optical-quality glass to be used for periscope lenses and gunsights. The mine reopened in 2012 and today 100,000 tons of glass sand are mined each year.

Once we assemble all the evidence the story becomes clear, that chalk formerly covered everywhere in the British Isles, except the Scottish Highlands, which were low-lying islands of crystalline granites, schists, and ancient sandstones, fringed with white sand beaches, much like today's Seychelles.

⋆

In Hampshire and Sussex the chalk layer is 560 metres thick, but the sequence runs out before we get to the youngest Cretaceous chalk layers. Younger chalks outcrop in the cliffs of northern Norfolk, yet even this chalk does not reach all the way to the end of the Cretaceous period.

To find what happened next we can travel to Denmark. Above the youngest Cretaceous chalk lies another layer of chalk termed 'Danian' (for Danish). Until the 1950s geologists considered the Danian to be a continuation of the Cretaceous—because Danian chalk looked much like the chalk beneath it. Then it was found that a global mass extinction had occurred immediately before the Danian, so the formation was relocated to the earliest part of the Paleocene, i.e. in the Cenozoic era.

Between the Cretaceous and the Paleocene, amazing things happened, things that were literally out of this world. Yet the record of this boundary is completely missing across the British Isles.

Denmark turns out to be the only place in northern Europe where the boundary can be visited.[13] Stevns Klint[14] cliff is the world's longest exposed section of the 66-million-year-old Cretaceous–Paleocene boundary.

The 40-metre-high cliffs run for 25 kilometres along the east coast of Sjaelland, facing the southernmost tip of Sweden. Less than an hour's drive from the international airport, this 'cosmic' destination is not hard to reach. Park the car and it is barely a minute's walk to the cliffs.

Twenty years ago, long sections of this coastline lay out of bounds in the Stevnsfort military complex. Armed with a monster gun removed from a German battleship, the fort protected the entrance to the Øresund and the capital Copenhagen while monitoring ship movements in and out of the Baltic.

Midway along the cliffline, the thirteenth-century Højerup Church, built out of chalk rock, stands on the lip of the cliff. The eastern half of the nave was lost in a rockfall in 1928 and the church now ends in a balcony overlooking the sea. Steps from the old church descend steeply down to the beach past the thin layer of clay that marks the Cretaceous–Paleocene boundary. The Danian chalk limestones are harder than the underlying Cretaceous chalk so the cliffs tend to overhang. Between the white Cretaceous and Paleocene chalks lies a thin black layer: the 'boundary clay'.

In 1980 Luis Alvarez (a Nobel prize-winner in Physics) and his son Walter first proposed that a massive asteroid had collided with Earth to cause the 66-million-year-old Cretaceous–Paleocene boundary.[15] In support of their theory they pointed to high concentrations of the extremely rare element iridium in sediments deposited at this time. In ordinary terrestrial marine sediments iridium is scarcer than gold, at concentrations of 1 part per billion. In a chemically differentiated planet (such as Earth) iridium joins the iron at the core. However, in an asteroid composed of the primordial material of the solar system, which never formed part of a planet, iridium levels are typically between 400 and 850 parts per billion. In the boundary clay from Stevns Klint, iridium reaches concentrations of 100 parts per billion. The sediments were also found to be rich in tiny shards of 'shocked quartz', which had been exposed to ultra-high pressures.

The 'smoking gun' for the asteroid collision was later discovered close to the north-west corner of the Yucatan peninsula, in south-east Mexico. The Chicxulub Crater is 180 kilometres across, part under the land and part beneath the sea. It was carved out by an asteroid estimated to have been 12 kilometres in diameter, with a relative velocity of 20 kilometres per second, arriving at a low angle, blasting material principally to the north, towards Haiti and North America. It is estimated up to half the mass of the asteroid was atomized and carried on the winds.

The biological activity that created the chalk came to a sudden halt. The boundary clay begins with the 'Fish Clay' (named for the high concentration of fish teeth and scales) and has three distinct layers. Starting at the base there is a 3-centimetre black to grey marl (lime-rich clay). The iridium concentration is at its highest just into the lowermost marl layer and then tails off upwards. Uniquely, 6–30 per cent of the materials in this basal clay are extraterrestrial. The first half-centimetre of the black clay contains iron-rich micro-spherules—vaporized and resolidified meteorite fragments. The sea was starved of oxygen, as the rivers and seas had become a vast animal and plant mortuary. Above the top marl, we can see that life slowly returned to the oceans, as the earliest Cenozoic Danian limestone began to form.

Along the most accessible parts of the cliff, the boundary clay has been worn away by 'apocalypse pilgrims' who rub their fingers and scrape their knives along the boundary, which of course is a visual disappointment. How can anything so catastrophic look so banal?

*

The spread of the chalk reveals that almost all the British Isles was underwater from 100 million years ago until at least 70 million years ago.

Meanwhile, deep within the Earth's mantle, hundreds of kilometres down, a vast plume of super-hot mantle material was slowly rising beneath the area to the north-west of the British Isles. The mantle rock was close to melting and moved like a viscous, sluggish liquid, less dense than its surroundings, rising inexorably higher and higher until, about 100 kilometres down, it began to spread out sideways at temperatures as high as 1,700°C,[16] 400°C higher than the ambient temperature at this depth. Imagine a plume like the molten wax column in a lava lamp.

The plume began to melt, the melt being less dense than the typical mantle rock. Some of this melt emerged at the surface, but even more of it pooled and solidified around the base of the 30-kilometre-thick continental crust. This is why Scotland, Ireland, Wales, and most of England exist as land today.[17] Before the plume, most of this region was below sea level. Afterwards, the British Isles was mountainous upland. The plume left permanent uplift as if buoyancy bags were placed under a ship.

The dome under Britain was probably oval in shape, elongated north–south, everywhere exposing chalk. 'This precious stone, set in a silver sea' is the landmass that eventually, 60 million years after it first emerged, would become the British and Irish Isles. The answer to the jingoistic rhetorical question in the words of 'Rule Britannia' as to when Britain first 'arose from out the azure main', is 'about 60 million years ago'. This was accomplished less 'by heaven's command' and more by a mantle plume. Today less than 10 per cent of the perimeter of the British Isles exposes the chalk. Yet at its birth, there were dazzling white cliffs of chalk continuing for 1,000 kilometres to the east and another 1,000 kilometres to the west of this great promontory, newly emerging out of the sea. The covering of chalk has been peeling back ever since.

In the 'bookshelf' geological section at Alum Bay, spanning the Cretaceous–Paleocene boundary, some 'pages' of prehistory either have been torn out, or were never recorded.

Since the beginning of the Cretaceous, 140 million years ago, the opening Atlantic separated North and South America from Europe and Africa. The asteroid impact was at the south-west end of this ocean, and the rising British landmass at the north-east end. The asteroid landed at the edge of the Atlantic, where modelling shows it generated a tsunami a mile high. When the tsunami reached what is now the west coast of Scotland and Ireland the wave may have towered many tens, even hundreds of metres high (Fig. 12). Did the tsunami

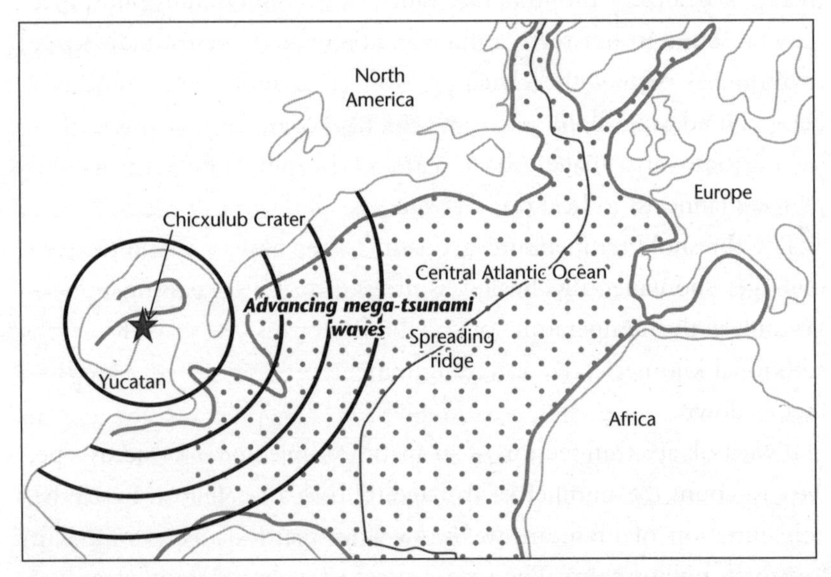

Figure 12. Chicxulub (66 Ma) asteroid impact and the contemporary North Atlantic

break into the North Sea, or was the shallow sea set in vigorous commotion by the global shockwaves? In the middle of the North Sea[18] borehole cores reveal a 60-metre-thick layer of redeposited, slumped chalk, situated immediately underneath the Fish Clay (with its iridium) laid down at the boundary. At the Stevns Klint section in Denmark, there are mounds in the boundary sea floor up to 500 metres long and 10 metres high, caused by strong water movements. In the Moray Firth there is deep erosional channelling at the boundary, caused by gigantic waves.[19]

*

Geological history has a powerful bias. When the land subsides or sea levels rise, sediments are laid down, recording the climate, the water depth, even the animals that lived at the time. But when the land rises and erosion kicks in, the record becomes much more tenuous.

Drill deep underground and the rocks grow warmer. As mountains rise and get worn down by erosion, the rock beneath cools because heat escapes to the surface. Cooling can have interesting consequences. In crystal grains of the mineral apatite (the same calcium phosphate mineral found in our teeth and bones) there are traces of radioactive uranium. As part of the chain-reaction of uranium-238 radioactive decay, blundering alpha particles smash into the crystal framework leaving trails of damage. Above a threshold temperature the crystal regrows (anneals) to heal this damage.

The threshold temperature in apatite, when alpha particle damage trails get spontaneously healed, is around 70°C. In passing underground, if the temperature of the Earth warms 25–30°C for each additional kilometre, the annealing temperature is reached 2–3 kilometres down.

If we look at a thin section of an apatite sample under a microscope, we can count the number of damage trails in a small area. From the concentration of uranium, we know how many damage trails will form in a million years. Then we can calculate how long that crystal grain has been below the 70°C threshold temperature.

By collecting samples of apatite from nearby rocks, but at very different elevations, for example, one from high up a mountain, one from the rocky shoreline, another from deep down a sea-level borehole, we find when each of these rocks passed below 70°C. If the mountain sample cooled through 70°C 60 million years ago, the sea-level sample at 55 million years, the deep borehole sample at 50 million years, then we know there was some rapid uplift and erosion 50–60 million years ago.

Counting damage trails in a section viewed down a microscope is laborious—the kind of work that PhD students find themselves compelled to undertake. (You only hope they have some good music on their headphones.) However, the results are not just of academic interest. The information is catnip to oil and gas explorers. Oil is generated by slowly simmering organic material in marine sediments. If conditions grow too hot, the oil will turn to gas and, like any organic material cooked too long, eventually reduce to carbon. If it never gets hot enough, the organic debris in the sediments won't have cooked to produce oil. Therefore, fission track studies (as they are called) help reveal when a source rock was 'in the temperature sweet spot' for oil generation.

In travelling to the north-west across England the fission track ages all point to uplift in the Paleocene, with the crest of this dome in the Irish Sea. Sixty million years ago this was an upland plateau, maybe 1,000 metres high, from which an estimated 2,000 metres of chalk and Jurassic sediments have since been eroded.

Why does the elevation of a 1-kilometre-high plateau followed by 2 kilometres of erosion not leave a deep sea? The crust of the continents, rich in lower-density granite, floats on the Earth's denser mantle, like a ship floats in the sea. If this was a fully laden container ship, imagine removing the top layer of containers. With a reduced load the ship now rides higher in the water. As a result, when measured as height above the waterline, the new upper level of containers will not simply be one container lower than it was before. It will be one container in thickness minus how much higher the ship is now riding in the water.

Very much the same applies to the erosion of uplands. As erosion proceeds and the load of rock is washed away, so the continental crust adjusts to ride higher in the mantle. This principle, also followed by floating icebergs, is known as 'isostasy'. A pillow of lower-density material now underlies western Britain, surmised to be up to 3 kilometres of solidified basaltic melt.[20]

Around Scotland there is evidence for both transient and permanent land-level changes. Fission track dates from borehole samples across the Moray Firth show that the region underwent a transient rise of from 180 to 425 metres in the Paleocene. By 55 Ma the uplift had reversed.

The geologist Brian Lovell, researcher and former BP exploration manager, showed the underlying mantle plume 'surging'.[21] In the Faroe–Shetland Basin land rose by 500 metres and then sank back again over 2 million years. When, around a million years later, the head of the plume had shifted 400 kilometres to the east beneath the northern North Sea, the transient uplift had reduced to 300 metres. Lovell saw such land-level oscillations as explaining rhythmic patterns common in the sedimentary record.[22]

And then there were also the permanent level changes, a tilt imposed across the Moray Firth basin that has led to 1,300 metres of uplift and erosion in the west, while zero uplift is reached on the lip of the North Sea basin.[23]

Across the new landmass, upland streams carved valleys and dissolved subterranean caves deep into the chalk. The caves linked into gorges, and the surface streams joined to form rivers radiating away from the centre of the dome.

Once it has carved a valley, it takes a lot of tectonics to displace a powerful river. The Colorado River in Arizona was not going to be shifted just because the land had slowly risen by a mile across its path. Instead it would just have to carve a deeper canyon.

The upper River Severn and the Wye flow to the south-east from near the crest of the dome, out beyond western Wales. These headwaters once fed the upper Thames, flowing out into the North Sea. The River Trent also rises to the west and flows east to the North Sea.

In Scotland the biggest rivers like the Forth, Tay, and Spey all radiate away from the dome that was situated out to the west. The River Shannon in Ireland flows in the opposite direction, away from the dome towards the south-west.

There were once other great rivers, now lost to marine erosion, including a river that ran to the south through Cardigan Bay between Ireland and Wales.

*

For hundreds of millions of years, the region that was to become England and Wales had been slowly sinking and accumulating sediments. And then, as we have seen, 60 million years ago the whole landmass was raised and eroded, exposing older rocks towards the centre of the dome. From London to the north-west, with only some modest gaps, the geological record that became exposed is continuous from the Cretaceous back to the origins of multicellular fossil life-forms, for almost 500 million years. This 400-kilometre journey provides a unique stairway into the past.

Yet it would take exceptional vision to see that a local story of preserved geological history had global significance. Step forward the self-funded autodidact, ex-soldier, and sometime country gentleman, Roderick Impey Murchison.

Murchison was a proud aristocratic Scot, born in 1792 in a mansion on the coast of the Moray Firth in northern Scotland. At the age of sixteen he served as a soldier in a number of Peninsular campaigns, but sat out the battle of Waterloo while billeted in Ireland. He lost his commission as the military was downsized following the defeat of Napoleon. In 1815 he married a talented heiress, Charlotte Hugonin, who had a deep interest in observing and drawing the natural world. After two years living in Italy they returned to Durham where Murchison sustained the indolent life of a country squire, riding to hounds and shooting pheasant.

He soon came to find this world stifling and, encouraged by his wife and a neighbour, the distinguished chemist Sir Humphry Davy, set out to pursue a more intellectual career. Loving the outdoors, he chose to commit his path to geology. He relocated to a fashionable central London address to be close to scientific club life.

Although he had no previous training in geology, in 1825 Murchison joined the Geological Society. A year later he was appointed one of the society's secretaries. To learn about this new science, he charmed his way into accompanying the most experienced geologists into the field.

First in his sights was the canal engineer and pioneer geological map-maker, William Smith, who had not been invited to join the gentlemen diners at the Geological Society. Murchison made overtures to 'Strata Smith' (as he had become known), as part of his own geological education. In 1831 Murchison branded Smith the 'Father of English Geology', and nominated him to receive the highest honour of the Geological Society—the Wollaston Medal. (The Geological Society's founding president, George Greenough, had earlier plagiarized Smith's geological map of England and Wales.)

But the geologist whose company and depth of experience he most sought was Adam Sedgwick, son of a country rector in Dent, west Yorkshire, and now Professor of Geology at Cambridge University. Murchison befriended Sedgwick and accompanied him on expeditions to the Scottish Highlands and the Alps.

In the 1820s and 1830s continental geologists started to assign generic names to the epochs of the Mesozoic geological strata.

The name of the youngest Mesozoic sediments, the 'Terrain Cretace', or 'Cretaceous', was coined by a Belgian geologist Jean d'Omalius d'Halloy in 1822. The next oldest, the 'Jurassic', named after the French Jura Mountains, was introduced in 1795, and in 1837 the leading German geologist, Leopold von Buch, defined the 'Jurassic' period.

Underlying the Jurassic was the 'Triassic'—meaning 'threefold'—coined by German geologists in 1834 for the characteristic trinity of Buntsandstein (Bunter sandstone), Muschelkalk (limestone), and Keuper (clays and salt formations).

The British were in danger of missing out on this opportunity to colonize geological time.

Across much of central Europe, between the Primary granites and gneisses and the Mesozoic strata, there were thick featureless sequences of mixed shales, breccias, and rough sandstones, known as 'Transition', or sometimes, from the German miners' term 'Grauwacke', greywacke.

The original idea was that the character of the strata made it possible to correlate sediments of the same age from one region to another. For example, 'Cretaceous' was French for the 'chalky' period. An older sequence of sediments was labelled 'Carboniferous' because they were 'coaly'. At the end of the eighteenth century, Abraham Werner, an influential Saxon mining geologist, claimed that different ages of rock could be identified by their appearance alone. Yet there were known to be problems with this theory. The Triassic rock formation known by German miners as 'Rothe todte liegende' (red sandstones) was initially considered the same age as the thick red sandstones that loomed over Hay-on-Wye, around the England–Wales border. But then it was realized that in Germany the sandstones lay on top of the distinctive coaliferous strata, whereas in England they lay beneath. The 'Old Red Sandstone' was much older than the 'New Red Sandstone' found in Germany and across the English Midlands, even though, in appearance, the rocks looked much the same.

Noting the parallels with the geology around Dent in north Yorkshire, where he had grown up, in the summer of 1831 Professor Sedgwick set out for north Wales to attempt to bring order to the Transition strata. In early August, at the recommendation of the Cambridge geologist and botanist John Stevens Henslow, Sedgwick called at the house of the Shrewsbury doctor, Robert Darwin, to meet up with his twenty-two-year-old, newly graduated Cambridge student son. Charles Darwin had abandoned medicine at Edinburgh and switched university to Cambridge, ostensibly to study theology, while pursuing his naturalist passions. He regretted not attending Sedgwick's geology lectures, but was now most keen to gain experience of field geology, in preparation for what he hoped would be a successful application to the post of naturalist/geologist on a 'round the world' royal naval survey.

Sedgwick chose to start his mapping in the north-west corner of Wales. Beginning with the oldest strata, he planned to advance slowly towards the east, to identify the fossils as he worked up the geological section. Charles Darwin travelled with Sedgwick to Bangor and Anglesey, learning to characterize rocks and identify fossils. He returned to his family home in Shrewsbury at the end of the month, where he found a letter from Captain Robert FitzRoy, reporting that his application to join the HMS *Beagle* had been successful.

In that same summer Murchison came to visit the Welsh borderlands, to see what had been reported a few years before: that in the river gorges of Shropshire the Old Red Sandstone layers continued down without interruption into underlying fossiliferous shales and limestones. While clambering around river cliffs identifying fossils, he had what he afterwards called his 'Eureka! moment'. He discovered he could extend Smith's fossil stratigraphy further back in time.

By June 1835 Murchison needed some smart naming to define this sequence of rocks and their characteristic fossils. He borrowed a poetic term already used to characterize the people of the Welsh borders: 'Silurian'.

The Silures were a valiant Celtic tribe who lived in the south-east of Wales when the Romans invaded, and who used their knowledge of the mountains and rivers to hold their own against the intruders, gaining the Romans' respect and eventually coexisting in their border towns. Through the eighteenth century the term 'Silurian' could be harnessed to describe the landscape of the Welsh borders, the proud inhabitants of the Marches, or even the local nobility. This was a brilliant piece of branding.

In response to Murchison's declaration of 'the Silurian', Sedgwick now coined a term for the lower end of the Transition formations in north-west Wales: 'the Cambrian'—the Latinized name for Wales.

The 1830s and 1840s were the peak of the railway construction boom in England. The search for coal was the primary utility of geology in the service of the industrial revolution. In the 1830s, Britain produced 57 per cent of the world's coal.

Murchison found there were no seams of coal or plant fossils in his Silurian formation, from which he inferred that land-based plants and trees did not exist at the time these strata were deposited. It would, as he frequently reminded his audiences, be a futile waste of money to search for coal seams in rocks of Silurian age—or older.

In August 1838, Murchison set out on a European tour, promoting his 'Silurian' formation, encouraging other geologists to adopt the term and inscribe it on their geological maps. Buoyed with new confidence, he wrestled with another geological riddle. How to interpret the rocks of Devon? In April 1839, Murchison proposed an intermediate formation, situated between the Carboniferous and the Silurian: the 'Devonian System'. Within three months Murchison was busy promoting the new name in France and Germany. A month after his announcement, the 'Devonian' was included in the latest edition of a popular geology book, Gideon Mantell's *The Wonders of Geology*.

As the master of the two newly coined brands, in April 1840 Murchison was invited on a grand tour of minerals and mining through Russia. Travelling 7,200 kilometres, while carefully noting the fossils, Murchison produced a map of the Silurian to the north, passing into Devonian and then Carboniferous formations to the south.

In the spring of 1841 he set out for a second phase of the Russian survey. From St Petersburg the party travelled to Moscow, and then on to Kazan, for the start of an expedition to the Urals. While travelling through Perm, the geologists recognized that a huge thickness of strata lay between the Carboniferous age rocks and the 'Zechstein' Magnesian Limestone found around the coasts of the North Sea, in Yorkshire and northern Germany. In honour of his Russian hosts he now proposed naming another geological system, sandwiched between the Carboniferous and the 'Triassic' New Red Sandstone—'the Permian', after the Perm region of Russia where the formation had been most prominent.

Murchison had now successfully named three of the principal geological periods as they became known—two from Britain, and one in Russia. No one else would name more than one.

Murchison was arrogant and imperious, amused when he was styled the 'King of Siluria', proud of his visits to meet the Czar of Russia. He was also author of a completely unappreciated British export. More than half the last 600 million years of Earth history is British-named, most of it a consequence of the 'Big Tilt' as interpreted by the ultimate imperial brand-builder, Roderick Impey Murchison. In 1847 Leopold von Buch remarked: 'God made the world, Sir Roderick arranged it.'[24]

One problem remained to be resolved. Murchison wanted to extend his Silurian to the very earliest fossil records, while Sedgwick sustained his Cambrian definition without being able to clearly define where it ended. This was not a quarrel they were ever able to resolve. Growing more and more ill-tempered in their positions and abusive in their conduct—Sedgwick accused Murchison of lying, Murchison claimed his rival was senile—by the 1850s they were no longer talking to each other and the Geological Society outlawed further debate. Sedgwick died in 1873 aged eighty-seven; Murchison predeceased him in 1871 aged seventy-nine.

It took the work of a new-generation geologist, Charles Lapworth, born in 1842 and fifty years their junior, to solve the puzzle. Trained as a teacher, he moved to the borderlands of Scotland, marrying the daughter of a local schoolmaster. In his spare time he painstakingly mapped the geology of the rounded mountains of southern Scotland. Instead of this being one great thickness of Silurian rocks, as previously believed, he showed it was a strongly tilted and repeated stack of shales and sandstones, characterized by tiny, stick-like fossils known as graptolites. The same fossils had been found in the disputed terrain between Murchison's 'Silurian' and Sedgwick's 'Cambrian' of north Wales. Lapworth resolved the argument about the Silurian–Cambrian boundary by proposing an intervening system. He only needed to find a suitable name.

The Ordovices tribe occupied the mid-part of north Wales, at the same time as the Silures were situated to their south. The Ordovices were most notable for their brave but suicidal attempt to resist the Roman invaders. Thereafter the Roman leader Agricola (according to the historian Tacitus) proceeded to exterminate the tribe. (They were widely considered 'losers' and no one subsequently used the term 'Ordovician' to describe the resolute locals, in the way they used 'Silurian'.) What chance that 1,800 years later, this savagely 'obliterated' and forgotten clan would be honoured by providing the name for 40 million years of prehistory, all around the world?

Starting from the origins of multicellular life, around 550 million years ago, the geological periods run: 'Cambrian' (named by Sedgwick, after being goaded by Murchison), 'Ordovician' (named by Lapworth to settle the unresolved quarrel between Sedgwick and Murchison), 'Silurian' (Murchison's first venture into naming geological periods), 'Devonian' (another Murchison special, supported by Sedgwick), 'Carboniferous' (tidied up and better defined by Murchison), followed by 'Permian' (yet again a Murchison original in honour of his grand Russian friends). That just leaves the Triassic (from Germany) and Jurassic and Cretaceous (from France), before we arrive at the Cenozoic, for which, of six lesser epochs (now grouped into two periods, the Palaeogene and Neogene), four (Eocene, Miocene, Pliocene, and Pleistocene) were named by Charles Lyell (Fig. 1).

A Chinese geologist, hammering samples from an outcrop in Egypt, will reference the age of the rock after a county in south-west England. Meanwhile, a prospector in eastern Peru identifies the formation she is exploring after a Welsh tribe, annihilated by the Romans almost 2,000 years ago.

★

It was one thing to assign names to the epochs of the past. Yet through the nineteenth century there was no way of knowing the actual ages of deep time.

The discovery of radioactivity, just before 1900, galvanized geology. The unstable element uranium undergoes radioactive decay and, after several radioactive elemental intermediaries, ends up as the stable element lead. If we know the rate of decay of the 'parent' element and can measure the concentrations of the radioactive 'parent' and 'daughter' elements it should be possible to date the rock.

Uranium–lead dating was pioneered by a young British geologist, Arthur Holmes. Born in 1890 near Newcastle-upon-Tyne, he came from a poor background, and struggled on a £60 scholarship to study Physics at the Royal College of Science in London (today's Imperial College). Mid-course he switched to Geology. On graduation, to earn some money, he became a mineral prospector in Mozambique, where he lasted six months, failed to find any gold or gems, contracted malaria, and learnt that his death had been reported back in England. On his return, he began doctoral research at his old department, on how to date uranium ores from the concentration of accompanying lead.[25] In 1913 Holmes published a book on the significance of radioactivity for geology titled *The Age of the Earth.* Arthur Holmes was the first to propose that it was heat, liberated by radioactive decay, that could drive mantle convection, sufficient to power the movement of the continents.

It was soon realized the dating situation was more complicated than first assumed. Different isotopes of the same element (same number of protons, different numbers of neutrons) could have quite different radioactive decay rates, while some isotopes were completely stable. Geological age dating required 'primordial' and radioactive isotopes, forged in the stars within which the material contained in planet Earth originated. Unless the half-life was hundreds or thousands of millions of years, the original radioactive parent would have decayed out of existence. Suitable for this purpose were uranium 238 (half life: 4.8 billion years; end member of a cascade of daughter products: lead 206), uranium 235 (half life: 700 million years; end member: lead 207), or potassium 40 (half life: 1.25 billion years; decays to argon 40 and calcium 40).

The measurement of tiny concentrations of parent and daughter isotopes had to await the mass spectrometer, developed in the 1940s as part of the programme to enrich uranium 235 to build an atomic bomb.

In dating the origin of some rock, care had to be taken to check the precious daughter isotopes had not already been flushed away due to heating or weathering, or, like argon gas, escaped through cracks. For this reason many isotope age measurements proved younger than reality.

Today, meticulous measurements of parent–daughter isotope concentrations have become widely used for dating pristine mineral grains and rocks.

Accurate isotope age dating was to prove critical to reconstructing what happened as the products of the hot mantle plume that had risen under western Britain burst through to the surface.

3

Age of fire

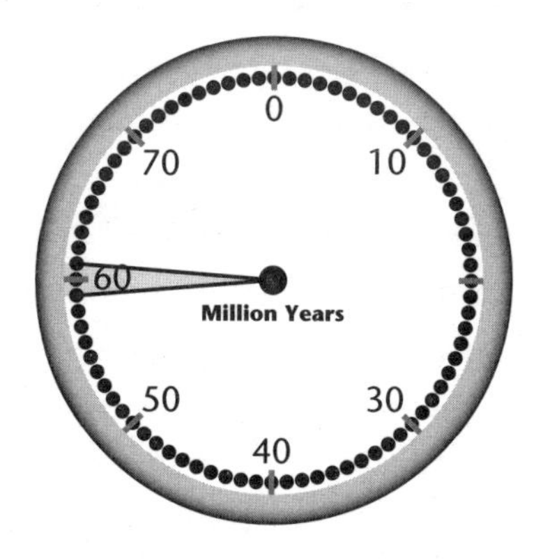

On their journey through the islands of the Hebrides in the stormy early autumn of 1773, Dr Samuel Johnson and his friend and chronicler Thomas Boswell paid scant attention to the landscape through which they travelled. Boswell disparaged the rocks of coastal Mull as 'black and gloomy'. 'Your country consists of two things: stone and water,' they told their hosts, reflecting: 'To the southern inhabitants of Scotland the state of the mountains and islands is equally unknown as with that of Borneo or Sumatra.'[1] Skye was considered the most romantic of the Hebrides islands: 'the Isle of Mists . . . full of the mystery of nature's charm and beauty, the spell of ancient and weird story'.[2]

In the 1790s Robert Jameson was the first geologist to explore the rocks of the rain-soaked Hebrides. In his 1800 *Mineralogy of the Scottish*

Isles Jameson wrote of Skye: 'this island appears, at some former period, to have been very much exposed to violent convulsions'. Schooled in the theory that basalt was somehow crystallized out of water, he did not recognize a past realm of fire.

Jameson lectured in Geology at the University in Edinburgh, where one of his audience was Charles Darwin.[3] Like Jameson three decades before, the seventeen-year-old Darwin was bunking off from studying medicine. Darwin considered Jameson's lecture style 'incredibly dull', so boring he would 'never attend to the subject of geology'. (However, in the spirit of student rebellion, he also assessed all but one of his lecturers in medicine as 'intolerably dull'.) As it happened, five years later, in order to gain his position on the *Beagle*, Darwin did 'attend' to the subject of geology, enjoying his crash course in field geology from Professor Sedgwick in north Wales. As a measure of the balance of Darwin's interests while on the *Beagle*, he compiled 1,383 pages of notes on geology, alongside 368 pages of biological observations. In a letter to his sister Catherine from the Falkland Islands, he admitted: 'There is nothing like geology; the pleasure of the first day's partridge shooting... cannot be compared to finding a fine group of fossil bones...'

When Darwin returned in 1836, Charles Lyell was president of the Geological Society, championing observations of land-level changes from the Chilean earthquake, even before the two had met. At the regular society meetings Darwin presented four geological papers, becoming society secretary from 1837 to 1841. Despite his 1837 admission of 'ignorance of English geology'[4] we will meet him several times through this story.

Skye's 'violent convulsions', we now know, were both volcanic and tectonic. The products of the oldest eruptions so far identified are found outcropping on the 'Small Isles'—Muck and Eigg.[5]

It seems likely that the landmass of Britain was doming just as the first fires broke through the ground, because they shared a common cause in the underlying plume.[6] The doming could not have proceeded too far because in parts of western Scotland, as on the Isle of

Mull, thin layers of easily eroded chalk have been preserved beneath the lava flows.

The next episode in the shaping of the landmass was igneous, 'born out of fire', introducing novel processes and materials from deep underground.

★

Magma is the name given to molten rock. The principles of magma chemistry can be learnt in the kitchen.

The melt

An impure solid melts at a lower temperature than a pure solid. The olive oil bottle was left in the freezer. Below −6°C the oil has turned into a waxy solid. Keeping both the butter and the olive oil frozen we chop them into grains and then consolidate them into a well-mixed mass, like the different mineral grains in an igneous rock. On a low heat, the mass starts to melt. Initially the liquid is almost all oil. As the temperature warms the butter melts (at 30°C) and the liquid is now a mixture, although still striped through with olive oil and butter. We find the same with magmas. As rocks start to melt, the first liquid will not be the same composition as when the whole rock melts. Instead, the first melt will be richer in everything that has a lower melting point.

Fractional crystallization

A Canadian winter is required to perform this experiment (but a freezer would also do). The goal is to make 'moonshine' alcohol.

Farmers pour apple cider into barrels and leave it in the freezing woods. As the cider gains a thick cap of ice, the farmer returns to break and remove the frozen slabs. Each time the liquid that is left, the 'moonshine', is richer in alcohol, because alcohol has a much lower freezing point (minus 114°C). With enough Arctic nights, and daily ice scooping, we can turn cider into calvados (apple brandy). The colder the winter, the stronger the brew. (But watch out you don't also enrich the hooch in toxic methanol.)

If a large volume of magma starts to cool, the first minerals to crystallize are those with the highest melting point. For a basaltic magma, the heavier magnesium-rich mineral 'olivine' crystallizes first as the magma cools, and sinks to the floor of the magma chamber, so what is left—the 'moonshine'—is increasingly rich in silica, as well as concentrating all the gases and water dissolved in the magma.

Deep heat

Pour hot chocolate sauce on solid ice cream and the ice cream will melt. The melting ice cream may slide off to the margins of the sundae. Only when stirred will the melted ice cream and chocolate sauce mix. Like hot chocolate sauce, a basaltic magma, originating in the mantle and flowing up cracks and channels into the crust, will melt the surrounding crustal rock (the 'ice cream') to form a granitic liquid, viscous and slow-flowing.

Beyond their chemistry, 'igneous' rocks (meaning rocks born out of fire) are named according to the size of their crystal grains. The faster the magma cools, the more crystals nucleate, bumping into one another and staying small.

Magma freezes quickly in a surface basalt lava flow, when the crystals are the size of table salt. ('Basalt' was in the past used as a dark, hard 'touchstone', on which the purity of gold could be tested.) At shallow depths underground, the basaltic magma chills more slowly. The crystals of 'dolerite' (from the Greek for 'deceitful', being easily misidentified) are the size of wheat grains. From the slowest cooling of all, in a magma chamber deep underground, we get 'gabbro' (a village in Tuscany), with large crystals, the size of wolf's teeth. If the magma is rich in silica, we get 'granite'.

*

Seen from the south coast of the island of Skye, when the moody rainclouds have parted, the island of Rum looks forbidding. Once a private deer-hunting estate, this is now Britain's largest nature reserve.

Each year the mountainous ruined stumps of the volcano provide mapping projects for midge-resistant geology students.

The magma tunnelled its way to the surface along a deep and ancient N–S fault zone—the Loch Long fissure, which still comes close to splitting the island in two. The volcano burst upwards from beneath the pre-existing geology, so that the surrounding sediment layers, on the edge of the volcano, are almost vertical. The volcano may have reached 3,000 metres high—perhaps the tallest mountain in the region for the last 100 million years.[7]

The jagged 'cuillin' summits in the southern part of Rum are the remains of a 'magma chamber', 500 metres across. The word 'chamber' gives the misleading impression that the walls and roof were somehow fixed, but the magma simply filled a gap, like water poured into a leather water pouch. Today the frozen magma forms dark, tiger-striped bands, rich in olivine and pyroxene crystals, interleaved with layers rich in white feldspars. As each fresh batch of magma started to freeze, the mush of higher-temperature, dark-coloured, heavier minerals sank to the bottom of the layer.[8]

The rounded summits to the west of Rum are formed from granite injected into the side of the former volcano. This was continental crust melted by the hot basalt magma: the 'ice cream' in the 'chocolate sundae'.

On one fateful occasion, superheated basaltic magma burst into a vast pool of cooler granite magma,[9] leading to a spectacular eruption of 10 cubic kilometres of devastating magma-froth, destroying everything around the volcano. As a final bow, the Rum volcanic edifice, having grown too tall and too steep, collapsed. Think of the collapse that launched the 1980 eruption of Mt St Helens, Oregon. Landslide deposits are peppered with boulders from all parts of the volcano, including fragments from the solidified magma chamber precisely dated to 60.33 million years.[10]

The Rum Volcano had been active for a few hundred thousand years. The whole volcanic episode in western Scotland only lasted 2–3 million years. In geological terms that is almost a blink of an eye.

★

As the fires went out at the Rum Volcano, a swarm of new volcanoes emerged on a pair of parallel lines (like volcanic 'ley lines'), oriented a few degrees anticlockwise from N–S (Fig. 13). The eastern of the two lines ran south from the Rum Volcano through the Ardnamurchan Peninsula Volcano, on to the Mull Volcano, and further south to the

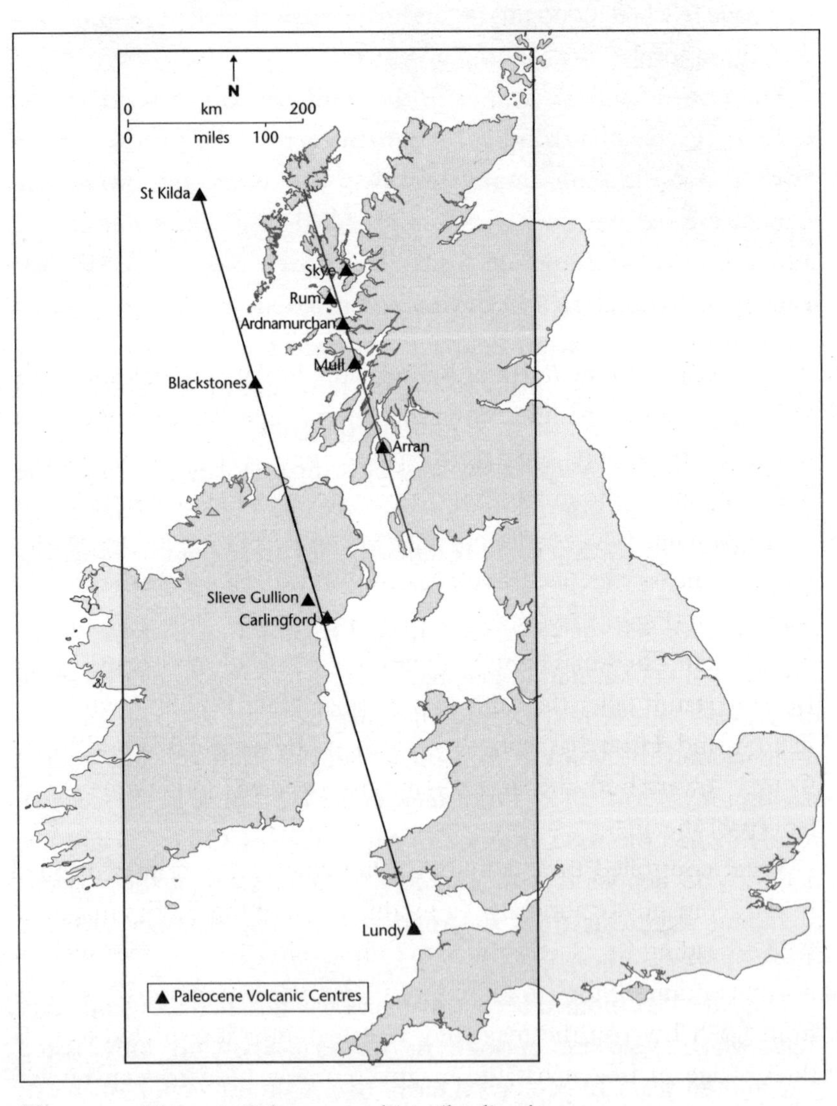

Figure 13. The twin Paleocene volcano 'ley lines'

Figure 14. Stacks of the St Kilda volcano archipelago

Arran Volcano. To the north of Rum lay the Isle of Skye Volcano, a few kilometres east of the line. Along the western line, the volcanoes were much more widely separated. Again there were five volcanoes. The remains of the northernmost volcano on the line comprise the St Kilda archipelago, 80 kilometres west of the Outer Hebrides, eroded into 'gothic skyscrapers' more daring and amazing than those of Manhattan (Fig. 14): two stacks as sheer as buildings, almost 200 metres high, lit like Christmas lights with nesting gannets. One soaring island, Boreray, stands taller than the 1931 Empire State Building, while the main island Hirta has cliffs higher than the original World Trade Center Towers; both the largest islands host ancient and unique populations of sheep.

What controlled these ley lines? One clue comes from a mapped linear geomagnetic anomaly—a modification of the Earth's magnetic field—heading in exactly this same direction a few degrees west of north, continuing for 110 kilometres across the Minch north of Skye, from Loch Ewe on the mainland. This feature is interpreted to be a thick ridge of frozen basaltic magma, entirely hidden deep underground, as wide as Edinburgh's Princes Street is long.[11]

The twin ley line rows of volcanoes were on the fringe of an enormous volcanic province, which extended from the western coast of Greenland to western Britain, all products of the underlying mantle plume.[12]

At this time Britain and Norway lay much closer to Greenland. The continents have since drifted apart with new oceanic crust laid down between them, formed at a mid-ocean spreading ridge. The widening continues today. There was not even a pebble of Iceland. Not a basalt lava flow of today's Faroes, parked next to the coast of Greenland. The Shetlands were closer to Greenland than to Aberdeen and the journey would have been walkable. However, 60 million years ago, Greenland was neither a continent of ice, nor even particularly mountainous, but an extension of the low-lying metamorphic rocks of the Canadian shield.

Today all these British and Irish volcanoes are in ruins, like ancient castles of which only some of the dungeons and foundations remain, or the stump of a tree that grew tall and then got felled by the wind and scorched by fire, only to grow back, over and over again.

An active volcano is inscrutable. All the internal plumbing is hidden. For a volcanologist, it is the ruined, eroded volcano that reveals its anatomical workings. In their exposed viscera, these Hebridean volcano stumps highlight four distinct processes.

The first, as at Rum, is a frozen magma chamber made from the primary basaltic magma liquid that flowed up from the mantle. A volcano may develop a succession of centres, as the original chamber freezes and new batches of magma arrive.

The second is what happens when the magma chamber rapidly empties, and the whole overlying volcano collapses to create a 'caldera' (the Spanish for cooking pot). Collapse leaves a steep-walled circular depression within a ring-shaped fault that dips away from the volcano centre. The gap between the 'piston' lid and the outer rim of the volcano can fill with magma to create a ring dyke, typically a few metres thick. (An igneous dyke is a near-vertical sheet of frozen magma that solidified underground.) The Loch Bà Ring Dyke on Mull records a massive collapse that produced a thick cylindrical dyke, 8 kilometres in diameter. The collapsed caldera typically fills with rainwater to become a crater lake.

The third element of volcano structure forms when the inner magma chamber is bursting with magma. The hydraulic pressure can lift the whole mountain. A cone-shaped gap is left filled with magma. Even after the volcano has eroded away, each surviving 'cone sheet' is testimony to the great mountain that once stood at this location, and the huge 'mountain-raising' pressures attained by the magma.

Last, granitic magmas, produced from melted continental crust, can be less dense than continental crust itself and they tend to rise as great pillows, many kilometres in diameter, up to the top of the pile.

The ruined Mull Volcano is a mash-up of dozens of ring dykes (each from a collapsed caldera) and cone sheets (each relieving over-pressured magma). On two occasions, as the primary Mull magma chamber became blocked, a new centre formed, a few kilometres to the north-west. Mull was once said to contain 'the most complicated igneous centre as yet accorded detailed examination anywhere in the world'.[13] Yet Mull is simply a volcano, exposed for dissection, whereas Skye is an even more complex mix of volcano and tectonics.

The basalt magma which flowed up from the mantle is more dense than the typical granitic rocks of the upper continental crust. Where this heavy magma has accumulated and frozen below a volcano, it increases the pull of surface gravity. Among these Hebridean volcanoes, the strongest gravitational force of all has been measured 60 kilometres WSW of the island of Mull. Under the sea lies the ruins of a volcano, contemporary with Skye and Mull, but now eroded and sunk completely underwater. The volcanic reef has been named 'Blackstones', after the basalt and gabbro rocks dredged by fishermen. The additional pull of gravity at the sea-surface is '152 milligals' (0.15 per cent of that from the whole Earth), the strongest pull of gravity around the British Isles. The submerged volcano has to be underlain by 5,000 cubic kilometres of dense mantle-derived rock, solidified from a repeatedly refilled magma chamber. If measured by the mass left behind, the mysterious Blackstones Volcano was the biggest of them all.

To the north of Blackstones, along the western 'volcanic ley line', the spectacular islands and giant stacks of St Kilda burst out of the wild Atlantic: as one visitor put it, 'a mad, imperfect God's hoard of all unnecessary lavish landscape luxuries he ever devised in his madness'.[14] The centre of the original volcano is defined by the accompanying 120-milligal gravity anomaly (second only to Blackstones), a few kilometres to the north of the main island of Hirta. Underneath lies a mantle-derived mass, 21 kilometres deep and 16 kilometres in diameter.

At Blackstones and St Kilda, what created the space filled by the huge volume of mantle derived magma? It seems inconceivable the same volume of continental crust was all simply melted away. Maybe this was a 'pull-apart' basin and the ley line was indeed a fault?

The comparison of gravitational pull could be misleading for indicating the volume of magma that arrived at each volcano. Skye and Mull, as we shall see, were principally transit stations, exporting most of their production.

The volcanoes lying along the western 'volcanic ley line' were not only much larger but also more widely separated than along the eastern line. These characteristics are likely related. The underlying 'magma catchment' (like a river catchment) was more extensive where the volcanoes were situated far apart. Only two of the western volcanoes, twins 'conjoined at their root', Slieve Gullion and Carlingford, are fully on land, even though separated by the Irish border. The other western volcanoes are little known, either by volcanologists or by the geologically inclined public.

★

At the same time the great volcanoes were active, basaltic lava flows poured over the surrounding landscape. On Skye any chalk had already been lost to erosion.[15] Around Mull, black lava rests on a layer of chalk, covering Jurassic age sandstones and clays. In Antrim the chalk was harder and the lavas flowed over a karst relief of caves, dolines, pinnacles, and gullies[16] with soils rich in weathered volcanic ash.

Britain was embarked on a slow drift to the north, having reached 40°N latitude—today's central Portugal. The climate was warm and moist. Volcanic ash rapidly degraded into rich soils sustaining thick forests. Sometimes the leaves and spores became washed into pools and fossilized, as at the Ardtun leaf caves on the west coast of Mull,[17] which reveal hazel, oak, plane, and gingko trees, along with magnolia and sequoia. Fringing the pools there were ferns and club mosses. At higher elevations the forest switched to pine.

Volcanic and tectonic activity combined with central volcanoes, lava fields, and lengthy planar igneous dykes. Thick piles of basalt lava flows have survived in three areas: northern Skye, north-western Mull, and the largest area, in Antrim, Northern Ireland.

The first magma to pour out on Mull buried the land in 5 metres of lava, a single eruption spreading over 20 square kilometres.[18] Production increased and after ten or twenty flows the lavas were 15 metres thick with each eruption covering hundreds of square kilometres. Before the lavas the land had been covered in mature forest. 'MacCulloch's Tree' on Mull, 12 metres tall and 150 centimetres in diameter, has been petrified by a lava flow.

Up to thirty of these lava flows are exposed in a ziggurat landscape on the Ardmeanach peninsula in western Mull. As the basalt surface weathered, a red soil would form. Occasionally a layer of peat became squashed and heated into a thin seam of coal. The layers are almost horizontal because the magma flowed like water, creating 'trap' landscapes (from the Swedish word 'trappa' for staircase) (Fig. 15).

We have an eyewitness account of such an eruption in Iceland, from the local vicar, the Reverend Jón Steingrímsson:[19]

> Around midmorning on Whitsun June 8th, 1783, in clear and calm weather, a black haze of sand appeared to the north of the mountains nearest the farms of the Sitha area of south central Iceland, so thick that it caused darkness indoors and coated the earth so that tracks could be seen.
>
> On the 12th the flood of lava spilled out of the canyon of the River Skafta and poured forth with frightening speed, crashing, roaring and thundering. When the molten lava ran into wetlands the explosions were as loud as if many cannon were fired at one time. At first this fiery

Figure 15. 'Trap' lava-flow landscape, western Mull

> flood followed the main course of the river, and then spread over the banks and out over the older lava fields which stretch out on both sides.
>
> On the 13th the... thunderings and great roarings now came from some distance to the northwest behind the mountains, with earthquakes and a constant humming and rushing sound, like that of a great waterfall or many bellows being blown at one time...

After pouring down one river canyon the lava appeared in a second, where it:

> flowed with the speed of a great river swollen with meltwater on a spring day. In the middle of the flood of fire great cliffs and slabs of rock were swept along, tumbling about like larger whales swimming, red-hot and glowing. When they struck something solid in their path or to the side of it, or if two of these great masses struck each other or were crushed together, they cast up such great sparks and bursts of flames hither and thither that it was terrifying to watch.

Steingrímsson waded through the flooding rivers and walked over the freshly chilled lava fields. When the clouds cleared he could see the fire fountains to the north, shooting hundreds of feet into the air.

Some 6 cubic kilometres of lava were erupted in the first week of the eruption, equal to two Olympic-size swimming pools every second. By the time the eruption was staunched, lava filled the 40-kilometre-long River Skaftá gorge. In total almost 15 cubic kilometres of magma erupted (twice the volume of water in Britain's largest lake, Loch Ness) spread over 580 square kilometres, in the largest basaltic eruption known from the past 1,000 years. The 1783 Laki Fissure eruption must have been very similar to all those unwitnessed eruptions that piled up the lava flows found along the west coast of Mull.

★

It would be hard to promote lava fields for their beauty. An honest reviewer would admit these are ugly masses of black slag, weathered a rusty brown. However, a small selection have become extraordinary tourist attractions.

In a 1693 letter to the Royal Society of London, Sir Richard Bulkeley, a fellow of Trinity College Dublin, announced the discovery of 'the Giant's Causeway' on the north-west coast of Antrim (Fig. 16a). The lava has chilled into some 40,000 hexagonal columns, a foot or two across, that form a three-dimensional architecture of steps and seats.

At the focal point of a grand amphitheatre of brooding basalt cliffs, an apron stage is thrust into the sea, a 100-metre, crocodile-shaped promontory on which the columns are the scales. There are no fences. The rock is so strong and the horizontal surfaces so rough that visitors are encouraged to climb onto the mass, recline on a columnar seat, strut their stuff, and strike a pose. Best visited early morning, before the crowds.

In 1739 the artist Susann Drury made paintings of the causeway, winning an award from the Royal Dublin Society. Through etchings that copied these paintings, the Giant's Causeway became famous across Europe.

The Causeway has had to survive the depredations of collectors. In 1759, Sir Hans Sloane let visitors to his newly opened British Museum, in Bloomsbury, enter over a floor made from Causeway columns.[20] Eventually a quarry had to be opened inland to satisfy the desire for visitors to take a column home with them.

Figure 16. Spontaneous geometry: (a) Giant's Causeway, and (b) Fingal's Cave

Equally famous is a small island to the west of Mull, named 'Staffa' by the Vikings because the encircling cliffs resembled vertical tree-trunks, as in their Stave churches.

A series of caves eat into the island's coastline. The largest measures 20 metres high and extends 75 metres into the cliff (Fig. 16b).

The original Gaelic name 'An Uamh Bhin' (the melodious cave) captured the restful echoes of the waves splashing against the columns. However, the cave was promoted as 'Fingal's Cave', first by Sir Joseph Banks, President of the Royal Society, who came to visit in August 1772, while on a journey to Iceland. He wrote:[21]

> The whole of that end of the island is supported by ranges of natural pillars, mostly above fifty feet high, standing in natural colonnades. Compared to this, what are the cathedrals or the palaces built by men! Mere models of playthings, imitations as diminutive as his works will always be when compared to those of nature.

The caves became a destination on the Celtic grand tour. Felix Mendelssohn was rowed into the cave in 1829. Four years later he published his orchestral sea-scape tone poem 'Fingal's Cave Overture'. In 1847 a boat decked with shawls brought Queen Victoria and Prince Albert.[22] Rudolf Steiner considered the cave a 'structure architecturally formed entirely out of the spiritual world itself'.

We now know how this spontaneous geometric architecture arose. Basalt magma ponded up to 40 metres thick in a river gorge, just as in the 1783 Iceland eruption. The stream that formerly ran at the base of the ravine was displaced to run over the cracked surface of the lava, chilling the magma and growing regular polygonal cracks.[23] Like the drought-cracks in a thick clay soil, the polygons tend towards the hexagonal, but there are columns with four, five, and seven sides. And thanks to the stream's quenching, the contraction cracks continued to grow right through the lava flow.[24] One day, extraordinary hexagonal columns may emerge from deep inside the thick lava in the River Skaftá gorge, grown in the months following the 1783 eruption.

In Skye the lava pile reached at least 1,500 metres thick.[25] At Mull 1,800 metres of lavas survived and another 20 per cent is projected.[26] In Antrim the lavas had a maximum thickness of only 1,060 metres[27] but covered a larger area than at the Hebridean lava basins.

The dates constrain the production of the Skye lavas to a period shorter than 1.6 million years.

*

In England a dyke is a drainage ditch. In Scotland a dyke is a dry stone wall bounding a field. In Holland a dyke is a wall to keep floods at bay. The word means a barrier which could be either a wall or a ditch. In its original Iron Age meaning the dyke was a boundary formed by digging out the material from a deep ditch to form a high bank. So the dyke was ditch and bank combined.

For geologists an igneous dyke is a big, near-vertical, opened crack now filled with frozen magma. The two sides of the dyke originally fitted tightly together. Opening happened at the same time as filling.

On the edges of big dykes there may be little branching dykes. Along the rocky foreshore in Connemara, there are dykes to be traced. A black band cuts the granite, sometimes splitting into two or three braided strands and then recombining, all the while maintaining the same total dilation, just like the crack in the mill-house wall.

When an igneous dyke slices across the landscape, sometimes for many kilometres, it can present both as a trough and as a raised causeway, sustaining the ambiguity of the original meaning, either wall or ditch. In most situations the igneous dyke occupies a ditch because the dyke-rock erodes more readily than the 'country rock' into which the dyke was intruded. In the presence of water, high-melting-point minerals, like olivine, rot and lose their strength and the rock disintegrates. On the granite island of Lundy in the Bristol Channel the dykes are almost invisible because they have been so deeply eroded into gullies.

Yet in some locations this situation is reversed. On the southern coast of Arran, a swarm of dykes form walls like breakwaters as they pass through weak sandstones. In the south-east corner of Great Cumbrae Island, a monumental upstanding dyke forms the Lion Rock.

Most spectacular is the west coast of the island of Jura, one of the wildest and most remote locations in the British Isles, a day's walk from any habitation along trails worn by deer and goats, and on a calm day best visited by kayak. As first identified by Robert Jameson in 1800, 'in some places the basalt decays first... in others the granular quartz decays first... leaving the basaltic rocks extending across the

Figure 17. Natural dyke walls along the west coast of Jura

beach like immense walls'. The shoreline dykes form extraordinary natural sculptures: a leaping animal, a set of houses, or just a massive writhing wall that one cannot believe is not a human fortification (Fig. 17).[28] For some reason the dykes are more resistant to erosion than the surrounding quartzites. In the channel between Scotland and Northern Ireland dyke walls up to 28 metres high have been spotted on the seabed.[29] (Is there some special ingredient in the dykes radiating from Blackstones?)

The dyke opens in the direction of extension, as determined by the tectonic stress-field. The path followed by the dyke reveals the stresses in the crust at the time it formed. The magma pressure cannot force the crust to widen unless the crust is already being stretched, across thousands of parallel cracks. When the magma arrives the opening transfers from the 'millefeuille' of open cracks into the single magma-filled dyke. The pressure of the magma is sufficient to force these thousands of pre-existing cracks to close.

Underground, the dyke forms on the weakest link, or the weakest crack, the one most prone to pull fully open when the rock gets stretched. Most of the time, and after the magma is fully frozen, the line of the dyke does not get reused for a second dyke. Once the opened crack has filled with solidified magma the dyke is no longer the weakest link. The two sides of the dyke are now firmly glued to the surrounding rock.

When the forces pressing horizontally are stronger than the weight of the overlying rock we have the conditions for creating the 'horizontal dyke' known as a 'sill'. The magma has to be under such pressure that it can lift up the load of overlying rock. As it solidifies the sill becomes a tabular mass, often 'staircasing' as it jumps from one stratal layer to another.

Sills misled early 'neptunist' geologists, like Darwin's first geology lecturer Robert Jameson. Neptunists believed basalt was somehow crystallized out of water. For a sill the layer of basaltic dolerite rock might be interposed between layers of sandstone, shales, and limestone. Proof, they thought, that dolerite was just another class of sediment. It took patient geological mapping to show how sills intruded, and postdated, the surrounding stratal layers, as well as the evidence that dolerite had been erupted from modern volcanoes.

★

Iceland lies above a hot rising plume in the mantle, just as western Scotland lay above a rising plume 60 million years ago. Deep beneath Iceland, as the hotter mantle material rises, it starts to melt, producing two or three times as much magma as at a typical mid-ocean spreading ridge. It is this productivity that has thickened the Icelandic crust and formed volcanic mountains, which rise more than 2 kilometres above sea level.

As a result, Iceland is the only place where the mid-Atlantic spreading ridge rises above the waves and can be visited on-land.

The oldest rock on Iceland dates from 13.4 million years ago, one-quarter the age of the Hebridean volcanoes. Individual lava flows in Iceland are much the same thickness as those in Mull. However, the pile of lava flows in Iceland is more than four times as thick as those in western Scotland.[30]

In Iceland today, just as in western Scotland 60 million years ago, the central volcanoes, dykes, and lava flows are all active at the same time. Did the magma first arrive up the dykes? Did the central volcanoes concentrate on the edge of the lava piles?

The volcanoes of the central Iceland spreading ridge are spaced on average 40 kilometres apart. Likewise the separation of the volcanoes at Rum, Ardnamurchan, and Mull. Arran is another 80 kilometres further south of Mull.

The largest Icelandic volcanoes at Askja, Grimsvotn, Bardarbunga, Katla, and Krafla have all developed 'calderas': circular, steep-walled, internal collapse structures, formed when the magma emptied out of the volcano. The Icelandic calderas can be hard to see because each volcano is capped with an ice sheet (Fig. 18).

There are two ways the magma stored inside a volcano can escape. In the classic explosive eruption, the magma bursts out the top of the

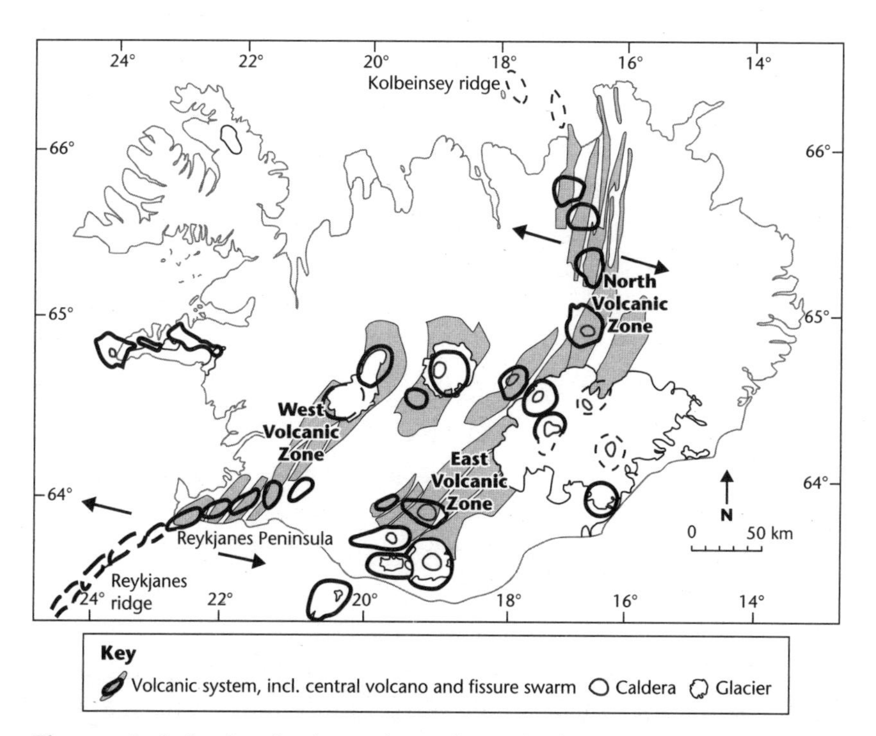

Figure 18. Iceland on-land spreading ridge and volcanoes

volcano. Worldwide, the last big, on-land, eruption was in 1991 at Pinatubo in the Philippines, leaving a caldera 2,500 metres across. On one occasion the Skye Volcano erupted a devastating mix of red-hot gases and magma, destroying everything in its path, like the Pinatubo eruption, out to 20 kilometres from the vent.

But there is another way for a basaltic volcano to empty its magma chamber.

The volcano Askja is situated in the middle of the main eastern rift zone in southern Iceland. In early 1874, towards the end of the winter, local farmers felt frequent earthquakes and saw steam pouring out of fissures. The volcano was preparing to erupt. Then between February and October 1875, at a location 50 kilometres north of the volcano, basaltic magma poured out from fissures and there was rifting. The lava chemistry was later found to be identical to that of Askja magma. The volcano must have fed the fissure eruptions through an underground dyke. As the magma emptied, the centre of the Askja Volcano collapsed, leaving a circular caldera.[31]

Even the greatest of all modern eruptions in Iceland, in 1783 at Laki, turns out to have been fed from a magma chamber, located beneath the neighbouring volcano, Grimsvotn.

In May 1783 local farmers noticed the ice sheet was melting and there were ash eruptions from Grimsvotn. Then in June the Laki fissure eruption began, vomiting 2,200 cubic metres of magma each second, sourced in a chain of fire fountains 50–60 kilometres southwest of the volcano. Over the following months, the magma drained out of the magma chamber and by the following February of 1784 the eruption was over, by which time the landscape to the south was buried under more than 12 cubic kilometres of new lavas. The caldera on Grimsvotn, hidden beneath the ice, measures 35 square kilometres, with near-vertical walls 600 metres high. When the 'cooking pot' was full, the volcano could hold 20 cubic kilometres of magma.

At the surface above the underground dyke that delivered the magma to the fire fountains, a 300-metre-wide and 8-metre-deep rifted trough had formed.

In north central Iceland, over five years starting in 1975, there were eleven eruptions at the volcano Krafla. The first was the largest. For a couple of hours an earthquake swarm migrated to the north, out to 50 kilometres from the volcano. The shocks were generated at the tip of the dyke as the magma forced the rocks to split open. At the surface, a 5-kilometre-wide rift developed.[32]

After 1975, the Krafla area was wired with monitoring devices.[33] By tracing the microtremors, scientists followed the tip of one widening dyke out to 30 kilometres from the volcano. For the first nine hours the dyke advanced at a slow walking pace, but on the following day reduced to a crawl. By measuring the deflation of the central volcano, the investigators calculated that the dyke had filled with 37 million cubic metres of magma, was half a metre wide, and had a vertical height of 2.4 kilometres.[34]

Beginning on 16 August 2014, a swarm of thousands of tiny earthquakes revealed a dyke propagating away from the central Iceland Bardabunga Volcano, between 5 and 10 kilometres underground. By 26 August the dyke was 40 kilometres long. Three days later a lava fountain eruption started another 2 kilometres from the volcano. By 1 September fire fountains were spewing more than 100 cubic metres of magma every second, while the volcano was sinking 40 centimetres each day. By November a full cubic kilometre of lava had been erupted, as activity sputtered to a halt.

In November 2023 a 25-kilometre-long underground dyke sliced through the town of Grindavik in the south-west corner of Iceland, splitting roads and houses apart, a prelude to a fissure eruption.

From the repeated experience of well-monitored Icelandic eruptions we can see exactly how the central volcanoes, igneous dykes, and lava flows fit together.

Perhaps against expectation, the dykes are not supplied vertically from underneath. In every one of the Icelandic eruptions, dykes were fed and propagated sideways, sourced from the magma chamber beneath a central volcano. Studies of the orientation of elongated crystals in British dykes confirm that the direction of flow was horizontal.

White-hot magma flows up from the mantle through a 'chimney' melted through the crust, perhaps tens or hundreds of metres across and tens of kilometres deep, supplying a central volcano. A few miles underground the magma pools and raises the roof of the geological formations to form a magma chamber, which becomes a holding station—like an old gasometer—in which the roof rises and falls according to how much magma is being held in store. Once the magma pressure gets too high, the sides of the volcano split open and the magma takes off on a widening vertical crack to fill an igneous dyke that radiates away from the volcano. If the pressure in the dyke is high enough, or the topography low enough, the magma in the dyke will break surface to feed a fissure eruption. And if it does not have sufficient pressure it will simply split the rocks at depth and continue on its way as a dyke, until it runs out of magma, or the magma cools and solidifies, blocking further horizontal flow.

At a few locations on Mull and Skye, as well as on Rathlin Island offshore from the northern coast of Antrim, it is possible to see where a dyke has fed an overlying lava flow.[35] Most of the time we simply have to infer that for every lava flow there was a magma-supplying dyke.

The dyke can only form if the crust is already extending, opening up sideways. For these Hebridean volcanoes, extension was limited to specific geographic locations, as defined by the paths of the dykes across Skye, Mull, and Antrim. When the magma chamber reservoir was full and pressurized but there was nowhere nearby extending, a sill could form instead of a dyke.

On the edge of the lava piles there are some big sills, like the 45- to 60-metre-thick Portrush Sill on the northern Antrim coast.[36] To the north of Skye, the Little Minch Sill Complex covers 4,000 square kilometres, with a total volume approaching 1,000 cubic kilometres. Minch sills make up the whole of the Shiant Islands.[37] A set of sills, in total 250 metres thick, are interposed in the Jurassic sediments on the Trotternish peninsula in north-east Skye, and along the eastern cliffs of Raasay Island. Underground there must be many more of these thick sills.

The youngest lavas to have survived on Skye are found at Talisker in the centre of the island, where they filled a canyon 120 metres deep (and named a single-malt whisky). The older lavas on Skye are compositionally more like ocean island basalts (formed through lesser degrees of melting of mantle rock).[38] The Talisker lava has a composition like mid-ocean spreading ridge basalts, formed from the direct melting of the plume. The oldest volcanic centre and magma chamber on Skye is now exposed in the Cuillin ridge, with the same mid-ocean spreading ridge composition, feeding dykes that supplied most of the original lava pile, now lost to erosion. Even then the magma is not a pure quintessence of the deep mantle. The geochemistry discloses, as with almost all these magmas, that en route to the surface, the superheated basaltic magma fused and then blended with molten continental crust.

The lava piles in Mull and Skye are located north-west of the big volcano which fed them. The lavas of Antrim look much the same as those in Skye or Mull, stacked one on top of another, only covering a larger area. Yet where was the central volcano that fed them?

At the centre of the typical volcano there is a vent where dissolved magma gases escape, continuously, or intermittently in eruptions. In an eruption, the magma tends to solidify close to the vent, accumulating a conical pile that builds up over many eruptions. When the magma chamber suddenly empties, the volcano collapses to leave a caldera. Conical mountain, central crater, and caldera—these are the geomorphic elements of a volcano.

It turns out that magma chambers have a 'sweet spot' for depth: between 6 and 10 kilometres underground.[39] Any shallower and gas emerges from solution in the magma and you get an 'uncorking the champagne bottle', runaway explosive eruption.

Were there one or more magma chambers beneath Northern Ireland without the apparatus of volcanoes? Across the eastern half of the Antrim lava plateau there are around thirty 'volcanic plugs', each the remains of a small volcano which erupted gas, ash, and lava in a single episode of eruption. Many of these volcanic plugs are oriented

NW and each was likely sourced from an underlying dyke.[40] On the north coast, one such plug is exposed at Carrick-a-Rede island, with the main attraction for visitors being a swaying rope suspension bridge. Huge boulders of chalk and lava are embedded in ash. The island is made of solidified basaltic magma that poured through the vent. Yet activity was short-lived. Even the stump of this volcano had been eroded before it was covered by later lava flows.

Farther west, a volcanic plug underlies the cliff-edge Dunluce Castle. The ash and boulders of a volcanic plug do not make for strong foundations. On an evening in 1639, without warning, the kitchen collapsed down the cliff, bearing with it the kitchen staff in the middle of preparing dinner. Their fate is not recorded. The tenants of the castle, considering this the final humiliation, moved out.

Each explosive vent, powered by magma gases, formed in association with a new dyke, while the dyke also fed an accompanying lava flow. That could explain the absence of a central volcanic crater.

The second mystery is, then, how the magma chamber deflated as the magma escaped to fill a new dyke, without creating an overlying caldera. We know some of these dykes fed massive lava flows, so there must have been magma chamber collapse.

So where was the magma chamber? Magnetic and gravity data tell us it may lie beneath the main 20-kilometre-wide concentration of dykes inland of Belfast, in the vicinity of Lisburn. Could the arc of Strangford Loch be a tilted section of caldera roof?

Gravity mapping suggests another basaltic magma chamber may be hidden beneath the Mourne Mountain granites[41] and another, perhaps again 20 kilometres in diameter, 3 kilometres beneath the NW-trending dykes of Fermanagh.[42]

In none of these places was there a topographic volcano.[43]

Mull and Skye look very much like 'Iceland in the Hebrides'—each a central volcano feeding a great pile of lavas through numerous dykes. Maybe the situation in Antrim was more like a typical spreading ridge, with magma chambers beneath the dyke swarms and no dominant volcanic centre.

4

The starter ocean

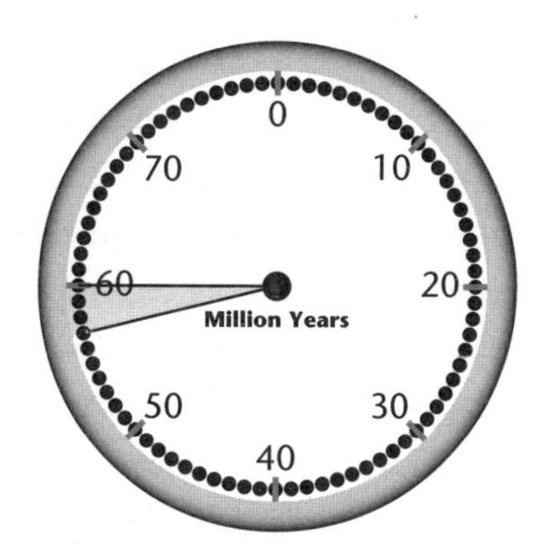

At the beginning of the 1960s an American geologist, Harry Hess, proposed that new ocean crust was being created symmetrically either side of 'mid-ocean spreading ridges'. Worldwide seismic monitoring had revealed a line of earthquakes along a submerged mid-ocean mountain chain that girdled the planet. In 1963 Fred Vine, a geophysics research student at Cambridge University, along with his supervisor Drummond Matthews, proposed that new basaltic crust, created in the centre of the ridge, became magnetized according to whether the Earth's magnetic pole pointed north or south. The magnetic pole switches on average once every million years, but randomly.

By 1966 their theory was confirmed. Measurements from ship-towed magnetometers revealed a mirror-image pattern of ridge-parallel magnetic stripes, like a barcode, around the spreading ridge. The farther away from the ridge, the older the ocean crust.

To explain how this magnetization worked required knowing the structure and composition of what underlies the ocean floor.

Wishing to capture some of the glamour of space exploration, in the early 1960s geologists devised the 'Moho Project': a 'moonshot' ambition to drill a hole through the crust of the oceans into the underlying mantle, an estimated 5 kilometres or more beneath the sea floor. It was soon realized you could save all the cost and the effort. In some locations, oceanic crust has become broken off and thrust up on land.

Even in Britain, sections of ocean crust and the underlying mantle have become embedded in ancient plate collisions: in Anglesey, on the Lizard Peninsula of Cornwall, and in the northern Shetland islands. The rocks of the mantle have reacted with water to create great masses of serpentine, a hydrated magnesium silicate mineral with the sheen of snakeskin and that breaks like industrial glass-slag. Serpentinite—the name of the rock—creates impoverished soils and has characteristic plants growing on it, like the Cornish heath with its long white and pink flower spikes. However, in the act of being caught up in past plate collisions, these British sections of ocean crust have become mangled.

Some of the best pristine and accessible oceanic crust can be found across the Hajar mountains in Oman, wrested from the bottom of the sea and thrust onto the south-east corner of the Arabian Peninsula 80 million years ago.[1] As you drive up into the mountains you pass through a complete section of ocean crust. At the base there is former mantle rock, known as peridotite, rich in olivine. Above that lies the ocean crust, starting with a few kilometres of coarse-grained 'basaltic' gabbro. Then come a kilometre or more of vertical sheeted dykes, overlain by another kilometre of lavas (Fig. 19).

Ocean crust, it turns out, has three layers. On the top are the *lavas*. These are fed by *dykes*, each sourced from underlying *magma chambers*. At the spreading ridge, every hundred years or so, the crust splits open

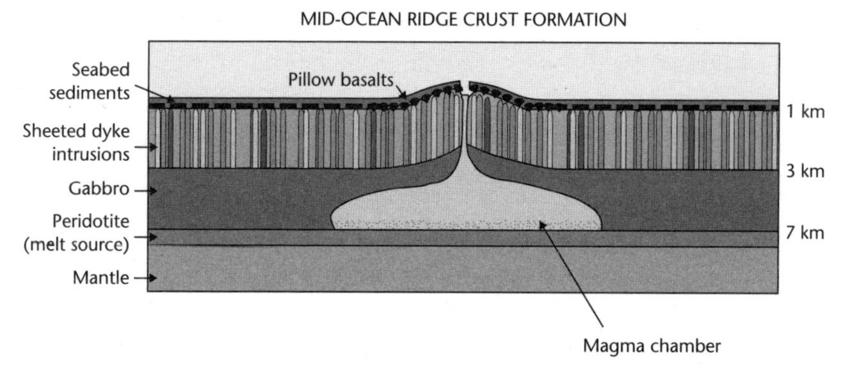

Figure 19. Oceanic crust

and a new dyke forms. The dykes intrude between one another so that the middle layer of ocean crust is made up of 100 per cent dyke.

Double (or treble) the thickness of each constituent of standard ocean crust (lavas, dykes, and magma chambers), and you have Iceland.

These three layers also exist in the Hebrides, but instead of each layer comprising 100 per cent of the crust, the basalt lavas, dykes, and magma chambers make up a fraction of the crust. The zone from Skye to Antrim was embarked on a path to become ocean crust, before abandoning the attempt. It is an almost unique example of a well-preserved 'starter ocean'.

How far had this starter ocean widened? How much progress had it made in the transformation into full ocean crust?

In 1819 the geologist John MacCulloch noticed the profusion of NW-trending dykes on the island of Skye, particularly well exposed along the rocky shore to the south of the Cuillin Hills. Over the past fifty years research students have spent happy days in wind and rain scrambling over the foreshore to measure the thickness of every dyke.[2]

Even with these valiant efforts, we can't count all the dykes. The shoreline has gaps in outcrop. In some places the lava flows continue below sea level, while dykes run underwater between the islands and mainland. However, extrapolating from what can be measured, sections where the foreshore is 10 per cent and even up to 25 per cent dyke, the Skye crust has been stretched and extended NE–SW by at least 3 kilometres.

One estimate of all the dykes, sills, magma chambers, and lavas on Skye adds up to 9 kilometres' thickness of frozen basaltic magma. Dividing by the 25- to 28-kilometre-thick crust, Skye could be said to be one-third 'oceanic'.[3]

South of the Arran Volcano, dykes expanded a 24-kilometre coastal section by 1,650 metres or 7 per cent.[4] The tally looks to be only half the stretching in Mull or Skye.

We know about the dykes of Mull, Skye, and Arran because these are islands with rocky shorelines. Across Northern Ireland, beneath the fields and forests, the dykes have only been made 'visible' through geomagnetic mapping. In 2005 and 2006, a high-resolution geomagnetic survey was made of the whole of Northern Ireland. To the surprise of local livestock (and farmers), the survey involved flying a De Havilland Twin Otter aircraft, bearing a magnetometer, at 56 metres above ground level (rising to 240 metres in urban areas), and then returning for the next traverse at a 200-metre horizontal separation. The dykes show up very clearly in the geomagnetic mapping because they are almost all 'reversely magnetized', their iron-rich minerals pointing towards a south magnetic pole.

Close to Belfast the crust has opened by 6 per cent, with twenty-five dykes in a kilometre.[5] Across Northern Ireland the spreading from the dykes spans a 100-kilometre-wide zone, and although the extension never achieves the percentage dilation of its Scottish cousins, it is likely the overall opening adds up to something similar (a little over 3 kilometres).

Spreading through shallow dyke intrusion thins the continental crust, drawing up the underlying mantle to fill the space. A reduction in pressure can cause the mantle rock to melt. Did the extending basins attract the volcanic activity? Or was it the volcanic activity that allowed each rifted basin to spread?

The initial volcano at Rum did not accompany significant dyke generation, so we can deduce that it did not take spreading to conjure up plume volcanoes. Meanwhile, contemporary with the dykes on

Skye, a 150-kilometre NNW-trending extensional fault opened to the east of the Shetlands, accumulating 800 metres of sands eroded from the uplifted Scotland–Shetland land mass.[6] With no underlying plume, extension could evidently happen without volcanoes. We can therefore deduce that the new spreading ridge plate boundary through the Hebrides and Antrim developed *because* the plume and its volcanoes had created a path of weakness.

Like the crack in the wall, the widening can't suddenly come to an end. The dykes on Skye must have been opening at the same time as the dykes on Mull, and those on Mull must have coincided with the dykes opening further to the south, across Arran and Antrim (Fig. 20).

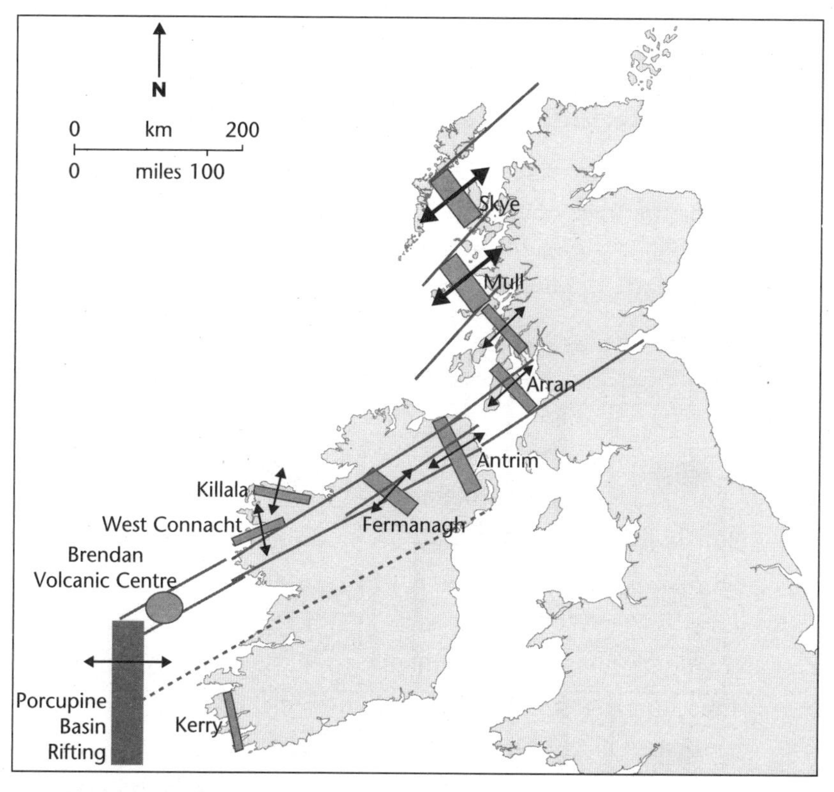

Figure 20. Paleocene dyke-swarm 'spreading ridges'

The 'starter ocean' was spreading sometime between 59 and 57.5 Ma. If spreading in the Hebrides proceeded at a similar speed to today's mid-Atlantic ridge (around a centimetre each year), a new metre-wide dyke would be intruded, on average, every hundred years. Three kilometres of spreading would take 300,000 years.

Similar oceanic spreading in continental crust is happening today beneath the rocky deserts of north-east Ethiopia, home to the Afar Rift. Just like at the Hebrides 59 million years ago, the dykes that feed the basaltic lavas are sourced from beneath central volcanoes.

★

In the late 1960s a Canadian geophysicist, John Tuzo Wilson, discovered a particular kind of strike-slip fault plate boundary situated between the offset sections of a spreading ridge. He called it a 'transform fault' (Fig. 21).

Transform faults have some special properties. First, they must align perfectly with the direction in which the plates are moving apart. Mid-ocean transform faults are only active, only generating earthquakes, in between the offset spreading ridge sections. Farther into the oceanic plates the line of the fault becomes a fossil transform, or fracture zone.

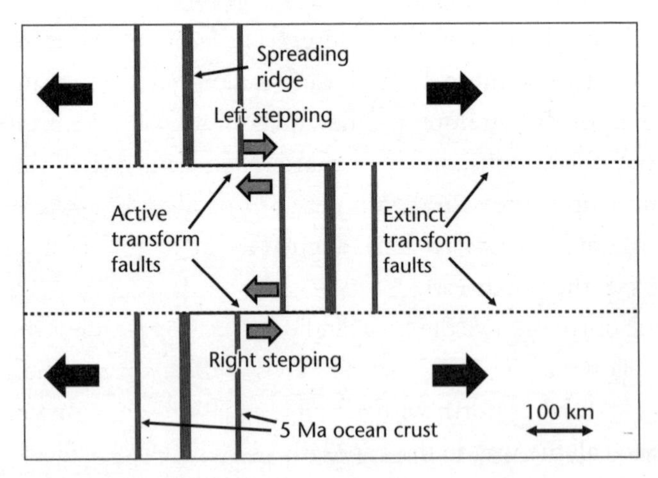

Figure 21. Transform faults between spreading ridge offsets

Although first identified along the mid-ocean spreading ridges, transforms can include other locations where the movement between the plates is carried by a strike-slip fault. The San Andreas Fault through California is itself a 1,200-kilometre-long transform fault.

On Iceland the spreading ridge splits. After arriving from the south-west up the Reykjanes peninsula, one path jumps to the east before heading north through central Iceland whereas the other runs through western Iceland. The central route appears to be replacing the western route.[7]

Because the rigid crust of Iceland is so hot and thin, instead of transform faults, the sections of spreading ridge simply overlap with one another. Only in passing offshore from northern Iceland does the spreading ridge revert to being offset by transform faults.

The pattern of dyke swarms and lavas in the Hebrides and Antrim[8] reveals that extension was segmented into three principal 'spreading ridges', each with an estimated 3 kilometres of opening. Were they interlinked via transform faults, as on a typical mid-ocean spreading ridge, or did they overlap with one another, as is happening in Iceland? In support of transform faults, the NE–SW Camasunary Fault links the Skye 'spreading ridge' with the Mull 'spreading ridge', while the south-west continuation of the Great Glen Fault[9] may provide part of a link from Mull to Antrim.

Like the crack in the wall, the offset sections of 'sea-floor spreading', from Skye to Northern Ireland, cannot simply come to an end. At either end of this attempt at a new plate boundary there must have been 'transform faults', to carry the displacement out of this region and link it into the worldwide network of plate boundaries. These transform faults must have been aligned in the direction of spreading, i.e. towards the north-east.

To the north of Skye the best candidate transform fault is the Minch Fault, following the east coast of Lewis and its continuation to the north-east. Farther north with a couple of jogs the opening may have transferred all the way to the Faroe Islands, where the earliest volcanic activity also involved swarms of NW–SE dykes.[10]

Such transform faults must have linked into the NE-trending plate boundary between Norway and Greenland, which was in sinistral strike-slip motion at this time.[11]

What happens to the south-west of the 'starter spreading ridge' in Northern Ireland? The geomagnetic mapping revealed three WSW-trending sinistral 'transform faults' cutting across the whole of Northern Ireland (see Fig. 20).[12] The younger the age of the dyke intersected, the smaller the displacement on these faults, reducing from 2.3 kilometres down to a few hundred metres. The faults were moving as transforms while the attempt at 'sea-floor spreading' proceeded.[13]

These transform faults are aiming for the sprawling Brendan volcanic complex[14] to the south-west of Ireland.[15] We can guess the rest of the path: eventually connecting[16] into the Charlie-Gibbs Fracture Zone—a key component of the North Atlantic spreading ridge architecture. We have followed the crack in the wall to its origins. The Hebridean 'spreading ridge' provided a temporary eastern diversion in the opening of the northern Atlantic (see Fig. 22).

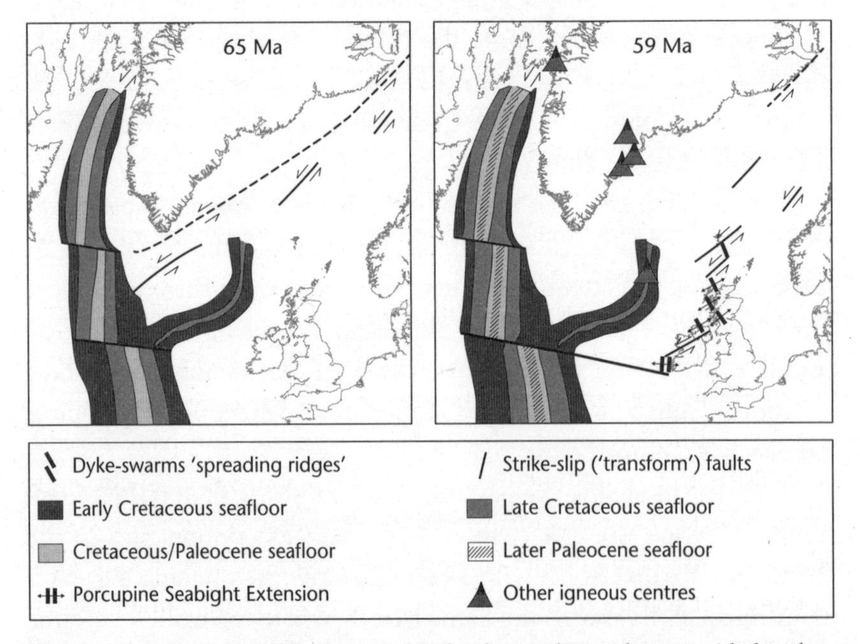

Figure 22. How the Hebridean and Ireland spreading ridges provided a plate boundary south-western detour

While the lavas on Skye and Mull poured out every few centuries without interruption, across Antrim for tens of thousands of years all eruptions ceased. In the heat and moisture, the exposed lavas rotted into a thick tropical soil.

Careful dating work reveals that the big central volcano at Slieve Gullion was erupting at the exact period when the Antrim lavas were interrupted.[17] This observation helps explain an incredible phenomenon.

★

At this time, something remarkable happened in north-west Britain and Ireland, something the physics for which is still not fully understood.

The dykes that fed the lava flows on Skye or Mull are identical to dykes in Iceland: typically a metre wide and 30–50 kilometres long.[18]

What happened next involved the production of dykes ten times as wide and ten times as long. Instead of opening metre-wide fissures that run for 40 kilometres we have dykes 10, 20, even 30 metres wide, some of which run for hundreds of kilometres. These are a different species to the ordinary lava flow feeding dykes. They are 'giant dykes'.

On the north side of the Mull volcanic centre the dykes exposed along the coastline are typically a metre or two wide. To the south of the centre, along the rocky foreshore, among the plentiful metre-wide dykes there are a small number of 10-metre-wide dykes. The giant dykes that converge on Mull first run SSE but then veer to the ESE across Scotland's Central Valley, continuing through the border country into north-east England.

I first learnt about the giant dykes from a promotion masterminded by the British Geological Survey in their two-sheet, second-edition *Geological Map of Britain* published in 1957. At a 1:625,000 scale even a 20-metre-thick dyke should only be an invisible 'thin-hair's breadth' one-thirtieth of a millimetre wide, but the mapmakers had chosen to treat the dykes like trunk roads and exaggerate their width almost a hundredfold. Straddling border country, the sweeping curves had become a dominant feature of the map. Then I discovered the same exaggeration of the thickness of these far-travelled dykes on a 'Geological, Railway and Canal Map of England and Wales' I have on the wall of

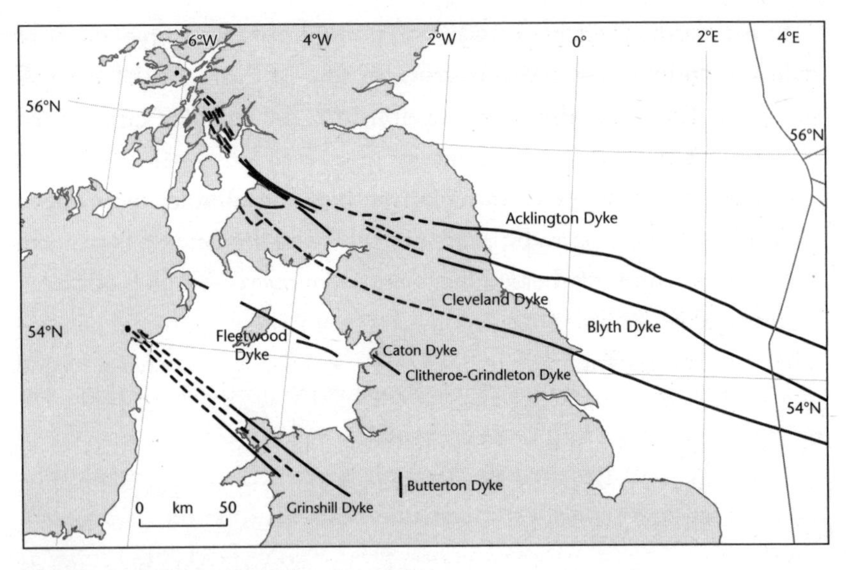

Figure 23. The paths of the giant dykes

my study, dating from 1843. Igneous dykes join canals, railways, and main roads as meriting special amplification, as though they were another class of artificial highway, paved with frozen magma.

The giant dykes of Northumbria have been lovingly documented and named[19] Moneyacres, Hawick–Acklington, Barrmill, Muirkirk–Hartfell, and Dalraith–Linburn (Fig. 23).[20]

One of these dykes is exposed in the rocky foreshore close to Tynemouth.[21] The Acklington Dyke cuts through the Cheviots and onto Durham,[22] passing through the former Togstone and Radcliffe collieries before moving offshore.

The most westerly member of this posse is the Cleveland–Armathwaite Dyke, which creates a cataract where it crosses the River Eden in Cumbria. The 28-metre-wide dyke continues into Yorkshire,[23] cutting the Jurassic hills at Great Ayrton. The dyke used to give a spine to the hills but the dyke-rock has been quarried to exhaustion to make cobbles for roads. In the 1880s, 8,000 tons of setts were transported each year by railway to Leeds.[24] After the excavation, all that is left of the surface trace of the dyke is a great big ditch.

From seismic reflection and magnetometer data, three of these dykes, including the Cleveland Dyke, can be traced far offshore, heading to the ESE.[25] As each dyke was intruded, outbursts of superheated water burst through to the contemporary seabed, creating a line of underwater thermal pools (now buried beneath younger sediments).

As observed on seismic profiles in the British sector of the North Sea, at least three of these dykes continue into the Dutch sector at around 54°N latitude, 600 kilometres from their source in Mull.[26] That is the end of the trail only because no one seems to have explored any farther. We simply don't know how far these dykes continued. Some geologists[27] blame the baking of coal seams in the Ruhr Graben in north-west Germany on the continuation of these dykes, by then over 1,000 kilometres from Mull.[28]

Mull was a 'giga-volcano', fed by an amazingly productive mantle magma chimney, with a magma chamber staging post, that supplied most of the giant dykes as well as the local lava fields. All around the volcano, the rock was stewed and metamorphosed by a vast pool of heat and fluid circulation. Many thermal springs must have emerged on the forested slopes, steaming on a calm winter morning.

In 1820, Cambridge geologist John Stevens Henslow (later tutor to Charles Darwin) discovered the largest of all the British giant dykes. On the muddy Anglesey shores of the Menai Straits, under a tangle of overhanging tree stumps, he located a dyke 40 metres wide.[29]

At the end of the nineteenth century, geologist Edward Greenly found that a continuation of this dyke exposed in a railway cutting at Capel Mawr in central Anglesey measured 53 metres across, perhaps a world-record width for any far-travelled dyke over the past 100 million years.

Henslow traced another giant Anglesey dyke from the south-east side of South Stack mountain in the north-west corner of the island, to cliff sections on either side of the sheltered beach of Porth Dafarch, where it is 24 metres wide (Fig. 24).[30]

Two of Anglesey's giant dykes outcrop above 750 metres elevation in the mountains of Snowdonia.[31] Another split dyke, 14.5 metres

Figure 24. Giant dyke encountered on the coast of north-west Anglesey

across, was encountered in tunnels at the Llanrwst lead mine, 20 kilometres south-east of Bangor.[32]

We think the next exposure of this dyke is 10 kilometres north-east of Shrewsbury, in the Grinshill stone quarry at Clive.[33] In 1834 Roderick Murchison identified the same dyke 4 kilometres away within the foundations for a grand house being built by one of his aristocratic 'silurian' friends at Acton Reynald.

The 12-metre-wide 'Fleetwood Dyke' outcrops on the north-east coast of the Isle of Man at Port Mooar.[34] From magnetic and seismic surveys the Fleetwood Dyke follows the line of an ESE fault across the Irish Sea towards Morecambe Bay, appearing as a series of parallel dykes and inclined sheets.[35] There is a rare instance of prospector's pathos in the tale of the borehole that cored 375 metres of Fleetwood Dyke dolerite when the seismic data had been interpreted to indicate a potential gas reservoir.[36] Despite its name, the dyke fails to make it all the way to coastal Fleetwood.

On the encouragement of Roderick Murchison, Charles Darwin became a dyke hunter.[37] In 1842, to escape from the vicissitudes of a young family, Darwin went for long walks around his wife's Wedgwood family home at Maer in north Staffordshire, 6 kilometres south of Newcastle-under-Lyme. On one such hike Darwin discovered an outcrop of a NNW-trending, 2.2-metre-wide dyke (now labelled the Butterton Dyke[38]). Magnetometer and outcrop mapping has today traced this dyke for 15 kilometres, from Keele University campus in the north to Norton Bridge in the south.[39] Unlike the dykes in Anglesey or the Scottish borders, for some sections there are two parallel dykes 130 metres apart, while in places the dyke has split into as many as five sheets.[40] Maybe splitting is a feature of a dyke reaching its terminus?

⋆

The giant dykes of Britain constitute an exclusive club. Six traverse the Scottish borders, but only three of them make it across the North Sea. One full-size dyke has been encountered in the northern Isle of Man, and a couple of more moderate dimensions to the south of the island. There are three impressive dykes in Anglesey, one of which makes it all the way to Shropshire with a couple more intersected down the Llŷn Peninsula and on Bardsey Island. And then there are (perhaps ten?) giant-width dykes in northern Ireland, including two or three up to 90 metres wide, and a few kilometres long, but apparently filled through several injections of magma.

The dykes of north-east England can be traced in a graceful ogee curve back to the Mull central volcano. This was a Scottish export to England (and for three of them, even on to Holland). The giant dykes on Anglesey appear to radiate from the Slieve Gullion Volcano (a northern Irish export to Wales). I discovered one giant dyke, usefully cleaned of vegetation and exposed like a surfacing whale, in a back garden, next to the coast road, 3 kilometres south of Carlingford town. My gawking was met with a hostile stare from the suspicious owner. These are Irish borderlands.

The Isle of Man dykes likely originated in a magma chamber beneath Antrim. Darwin's 'Butterton Dyke' presents its own mystery, 300 kilometres from the nearest volcano. The dyke is not a 'giant' by width, but simply by its having travelled a long way from any plausible source.

The orientation of a few degrees east of south is incompatible with all the other SE-trending giant dykes. (Intriguingly, this orientation is shared with the 'volcano ley lines'.) Different stress regimes could not have coexisted at the same location. Therefore, the Butterton Dyke must have been intruded at a different time to the other giant dykes. The orientation suggests it might be a little older than the other dykes, whereas the age dates, not always the most reliable, come out as younger: 52 Ma from potassium–argon dating and 54 Ma from fission track dating.[41] The chemistry of the Butterton Dyke is also different to the other long-distance giant dykes, classed as an alkali basalt with the characteristic low-silica mineral nepheline.

The Butterton Dyke has not been traced north of Keele. Projecting its alignment another 100 kilometres it would arrive at Clitheroe, Lancashire, where there is another mystery 'orphan dyke'.

The Caton–Grindleton dyke is geochemically similar to Butterton, an 'alkali basalt' (with nepheline),[42] first mapped at Caton 6 kilometres north-east of Lancaster where it trends at 150 degrees and is 1.5 metres thick, heading for its next identification, 25 kilometres away, as the Grindleton Dyke beneath Clitheroe.[43]

Perhaps the Caton Dyke was supplied from the Arran Volcano, 200 kilometres away (where some sills have a similar composition)?[44] Yet for Caton–Grindleton to pass into Butterton would require the dyke to turn a 25-degree corner, something that dykes typically don't accomplish because it would require a 'disjuncture' in the stress field.

Darwin was fated to discover the most enigmatic of all the long-distance dykes. And the strange destiny continues. Bizarrely, on account of the Darwin connection, a rock from the Butterton Dyke was sent into orbit on the Mir space station.

For all the giant dykes there remain unsolved questions. As with standard-size dykes, all the evidence suggests giant dykes filled sideways. The chemistry of the Northumbria dykes matches Mull magma,[45] including evidence for chemical differentiation in the original magma chamber. Yet how could a volcano hold enough magma to fill a giant dyke? A 20-metre-wide dyke, continuing for 400 kilometres with a depth of 10 kilometres, requires 80 cubic kilometres of magma, six times the volume of the 1783 Laki fissure eruption.

The bigger the diameter of the caldera, the greater the potential magma storage. At the Arran Volcano the caldera was 5 kilometres in diameter and at Mull 8 kilometres. For the largest of all the well-exposed volcanoes, at Slieve Gullion, the caldera measured 12 kilometres in diameter. A 1-kilometre rise of the caldera lid could theoretically store 100 cubic kilometres of magma.

The Mull and Slieve Gullion volcanoes must have filled with magma over thousands of years, becoming gravid with magma, bulging at the seams, swelling the whole mountain, until one day an underground crack opens and the magma takes off, and the crack continues to grow for hundreds of kilometres. It must then have taken thousands of years for the volcano to recharge. Yet why did the magma chamber not empty earlier to form a standard-size dyke? It seems that the maximum extension was not happening in the vicinity of the volcano.

For me the most amazing feature about these giant dykes concerns how the crust could store 40 metres of potential opening (or strain) before the dyke arrived.[46] Look on the mighty dyke exposed at Porth Dafarch close to Holyhead and be amazed that this formed in a single episode, over a few days, with magma supplied from almost 200 kilometres away in northern Ireland. The walls of the giant dyke did not widen by tens of metres from magma pressure alone. Imagine just the weight of the crust 10 kilometres deep along a length of 200 kilometres and maybe 100 kilometres either side of the dyke. That would require moving a hundred trillion (100,000,000,000,000) tons of rock.

The magma can only 'push against an open door'. The capacity to extend the crust by 40 metres must already have accumulated in the crust before the dyke was intruded: stored in the opening of millions of suitably oriented cracks in the crust. Over a 200-kilometre-wide zone, that would be 20 centimetres of opening per kilometre, or a water-filled crack open by the thickness of a credit card every 4 metres. As the magma arrives, all this opening transfers to the dyke, and the millions of cracks close, squeezing out the water and the gas.

Giant dykes are the equivalent of the very largest magnitude 9 earthquakes, like the 2011 Tohoku earthquake in Japan, which involved a fault displacement of more than 40 metres. Britain and Ireland's Paleocene 'Starter Ocean' volcanic province should be a UNESCO World Heritage Site and Anglesey's best exposed giant dykes a prominent tourist attraction.

Although they reached shallow depths of burial—there are frozen bubbles or 'vesicles' in the Cleveland Dyke rock, evidence that gas was coming out of solution, like uncorking a bottle of champagne—the giant dykes did not break surface and feed lava flows. If a dyke surfaces, and magma pours out, the magma pressure is relieved and the dyke won't grow any longer. The giant dykes that reached England and Wales grew and grew, so they must have remained entirely buried.

Or to put it another way, the volume of crustal opening and widening was greater than the amount of magma available to fill it.

In an open crack, as the pressure reduces, the top of the dyke will fall with increasing distance from the magma chamber. The dyke started off entirely underground, so where we find the dyke outcropping today depends on how much erosion has occurred over the past 60 million years. In particular when passing across Scotland's Midland Valley and the border 'peel tower country', many of the dykes have gaps in their outcrop. Several (but not all) shift alignment when they cut the Southern Upland Fault. Underground, each giant dyke must be continuous or the magma would not have been able to flow along it.

Unlike in Antrim,[47] there was no gap in lava production on Mull or Skye. The Mull Volcano was still supplying dykes and lava eruptions to the north of the volcano, while it was driving sporadic giant dykes to the south-east. Unlike the dykes that fed the lava flows, the giant dykes seem to have had no limit to their length, because the crust was extending across north-east England and the southern North Sea.

Rocks of the crust are subject to stresses, with lateral forces acting from the plate boundaries and from undertows in the mantle beneath. Where we find the profusion of dykes beneath the lava fields of Skye or Mull, or even with the giant dykes continuing into north-east England, there was active crustal extension under way. More generally, even without the extension, pump water down a borehole until the pressure of the water is greater than the minimum stress in the rock, and a thin water-filled crack will open. The weakest stress direction in the rock is at right angles to the crack. This is known as hydrofracture or 'fracking' for short. Ironically, the opening and closing of cracks across northern England accompanying the intrusion of the giant dykes seems to have flushed out a lot of the gas it was anticipated could be recovered from fracking.

*

Across north-west Ireland there are several splays of Paleocene dykes, and among them a number of wide dykes,[48] although none nearly as far travelled as those that make it to England and Wales (see Fig. 20). In County Fermanagh half a dozen dykes up to 30 metres wide have been identified from the geomagnetic mapping. The 90-metre-wide Doraville Dyke close to Loch Erne has been quarried for roadstone: a 3-kilometre-long 'multiple' intrusion that saw repeated arrivals of magma.[49]

On the coast of Donegal, most NW-trending dykes are standard metre or two sizes but there is also a lone 25-metre-wide dyke.

Across the Hebrides and into Antrim, most dykes run NNW—revealing the crust was extending WSW–ENE. In south-west Ireland a single 8-metre-wide NNW dyke can be followed for 75 kilometres from Smerwick Harbour to Black Ball Head reflecting the same stress field. Yet across north-west Ireland the dyke directions become less inhibited: NW–SE in County Fermanagh through to Donegal, E–W in the Killala dykes on the coast of County Mayo, and NE–SW for five West Connacht dykes in the cliffs of County Galway,[50] where the fattest is 22 metres thick.[51] These dyke swarms are much the same age and do not overlap. To the north-west of Ireland runs the deep-water Rockall Trough 'moat', beyond which there is the sunken Rockall Bank microcontinent. Before the Hebridean spreading ridge, there was NW–SE opening in the Rockall Trough and it seems the local stress field in the north-west corner of Ireland still came under this influence (see Fig. 22).

The NE–SW West Connacht dykes can be traced for at least 50 kilometres, the longest for 100 kilometres, and, like the Dingle Dyke, were probably supplied from the Brendan Centre: a vast (40 kilometres in diameter) volcanic complex 80 kilometres off the coast of western Ireland that is unexplored, unsampled, and unmapped. (The one geology of no interest to oil and gas explorers is a volcano.)

★

This extraordinary episode of giant dyke intrusion happened 59 million years ago. It was fed by a small number of Celtic volcanoes with their bellies full of magma, while Britain was being stretched from the north-east to the south-west. This new exploratory plate boundary only succeeded in widening by 100–200 metres as the 'giant' cracks became filled by magma. Then it gave up.

Now we can explain why lavas stopped pouring over Antrim as the giant dykes were being intruded. As long as the Mull Volcano was generating giant dykes towards the south-east, and the Slieve Gullion Volcano was feeding dykes to the south-east and north-west, the dykes beneath Antrim had no reason to extend and feed the Antrim lava flows. The crust was already opening farther to the east and west.

When the giant dyke 'experiment' was over, the spreading ridge under Antrim came back to life and lavas once again poured over the thick soils and jungle vegetation. At Tievebulliagh Hill on the edge of the Antrim Plateau, the bauxite-rich laterite soils became baked into black porcellanite, rich in corundum (aluminium silicates), perfect for the toughest Neolithic weapons.[52]

Beneath these volcanoes, towards the end of their activity, the continental crust had melted into huge pools of granite magma. As it solidified, great blobs of the low-density granite rose up through the crust to the top of the pile. The granite on northern Arran is 13 kilometres across and pushed up the surrounding rocks by 3,000 metres.

A smaller blob of granite magma rose up to become the island rock of Ailsa Craig, 20 kilometres south of Arran.[53] Cooling fast on account of its size, the rock was left as a fine-grained 'microgranite' tinged with blue or red minerals. The 200-metre-high 'sugarloaf' island is super-tough, resisting erosion from passing ice-sheets. In the late nineteenth century the same toughness meant Ailsa Craig microgranite became the principal source of curling stones. Today 'Kays of Scotland' harvests loose boulders off the island from which it carves and exports 17- to 20-kilo weight Ailsa Craig curling stones. The principal market is Canada[54] where 80 per cent of the world's 1.5 million curlers reside. Since 1998 all curling stones in the Winter Olympics have been sourced from Ailsa Craig.

★

What was in the back of Tolkien's mind when he imagined Mount Doom, the isolated volcano that anchors both *The Hobbit* and *The*

Lord of the Rings? Located in the north-west corner of the bleak land of Mordor, Mount Doom was the only location with mantle-sourced fires hot enough to both forge and destroy the magical ring.

Unknown to Tolkien when he was writing these mythic stories, since the emergence of the landmass that became the British Isles there was only ever one candidate volcano in the vicinity of Tolkien's English shires. Maybe reaching 2,000 metres high, visible in eruption from 100 kilometres to the east, here was the real Mount Doom.

Yet Lundy was always a most implausible volcano, more likely a work of fiction. Paleocene volcanoes 'happened' in the Hebrides, or in northern Ireland, not 20 kilometres offshore Devon. The island was only identified to be a Cenozoic volcano in the 1950s. Previously geologists assumed Lundy was the same age as the 280 million-year-old Dartmoor granite. The mineralogy and chemistry looked similar: they both formed through melting much the same continental crust. Yet uranium–lead isotope dating reveals that the Lundy granite solidified at 59.8–58.4 Ma, just like the volcanic centres at Skye and Mull.

Lundy is on the western 'volcanic ley line', a southern sister to the 'double plus'-size volcanoes at St Kilda, Blackstones, and Slieve Gullion. At the end of the line, more than 300 kilometres south of its nearest neighbour volcano, Lundy's magma supply was only a fraction of that of its northern sisters. Beneath the original volcano centre, 10 kilometres west-north-west of Lundy Island, gravity measurements show that basaltic magma intrusions only extend down to a depth of 4 kilometres.

A dyke swarm can be traced north-west for 30 kilometres, maybe formerly feeding lava fields. Two or three dykes have also been identified on the mainland, at extreme low water in Lee Bay, west of Ilfracombe.[55]

Once a 20-kilometre-radius mountain, today almost all that remains on land is a 5-kilometre-long granite monolith, like a giant tombstone, creating a plateau that rises in the south to 143 metres. Composed of melted continental crust, the granite solidified at least 10 kilometres underground[56] before rising close to the surface. What is now Lundy Island was originally on the more sheltered eastern side of the volcano, some 98 per cent of which has been removed by Atlantic storms.

The volcano was still active at 57.2 Ma when the granite was intruded by 200 basaltic dykes, up to 4 metres thick,[57] which radiate from a point 2 kilometres west of today's island, from what must have been a magma chamber beneath a later summit of the volcano. There was not much cover above the granite by this time and the dykes are full of frozen bubbles.

In the south-east corner of the island, close to the harbour, a fault separates the granite from metamorphosed rock described as greenschist. Between the island's 'Quarter' and 'Halfway' stone walls, subdividing the island plateau, an area of fissures is known as 'the Earthquake'.

Far enough from the mainland, surrounded by steep cliffs and in the middle of a vital shipping lane, Lundy was both an attractive refuge and a target for invaders. In the thirteenth century the island was occupied by despots challenging the English king. At other times it was attacked by raiders from France, Spain, and even Moorish north Africa.

The 1860s saw an attempt to open a sea-level super-quarry on Lundy. There was talk of supplying granite for the Thames Embankment in London. Money was raised to construct tramways, a harbour, and housing for 200 workers. However, rumours of skulduggery among the directors conspired with doubts about the granite's quality. By 1869 the company was in liquidation and the island was returned to the guillemots and puffins.

Lundy deserves to be a sacred space—this commanding remnant of the only English volcano in 100 million years, located at the intersection of a volcanic ley line and a great Bristol Channel fault.

Memorial stones record burials from the fifth to the eighth centuries, and a church has existed since the thirteenth century. Although this was never Iona, the volcano island has today become a secular pilgrimage destination, to view its spectacular western cliffs, numerous migrant birds, and breeding populations of puffins, razorbills, guillemots, fulmars, and many gulls.

There is still a church on Lundy. Dedicated to St Helen and opened in 1896, its suburban gothic revival style looks out of place and proved a target for Atlantic gales. With lottery funding the church has now become St Helen's Centre for study, education, and shelter from the storms, as well as the occasional wedding. With a ring of ten bells in the tower, the island receives many visits from groups of bell-ringers intent on a three-hour 'full peal', to the delight and distress of other island visitors.

Two or 3 million years after they first emerged, the ten great volcanoes of western Britain and Ireland emitted some final, modest eruptions, and then their fumaroles quit smoking, their hot springs cooled, and the surface of the last lavas turned into a thick red soil. Forests returned to cover all the lava fields and ash scree. As the volcanic activity declined, on Skye and Mull erosion and buoyancy brought the solidifying magma chambers up to the surface. Finally, each volcano was 'extinct'.

Today the island of Mull reveals the viscera of a very active volcano, which shifted its centre three times and is accompanied to the west by dyke-swarm-fed lava fields. In contrast, Skye was not only a great volcano, dyke swarm, and lava pile but has also been later rearranged by big tectonics.

Meanwhile, the mantle upwelling had moved to the north-west. An even greater cavalcade of volcanoes and fissure eruptions was about to be unleashed to the west of the Shetlands, and on the far side of the Rockall Bank microcontinent. The forces on the plates had shifted into a new configuration. First there had been the Hebridean to Antrim 'starter ocean' of dyke swarms and lava basins, then the giant dykes stretching deep into Wales and England. Now the third attempt at a spreading ridge would prove to be successful, finally unleashing Greenland from Europe.

5

Heat and dust

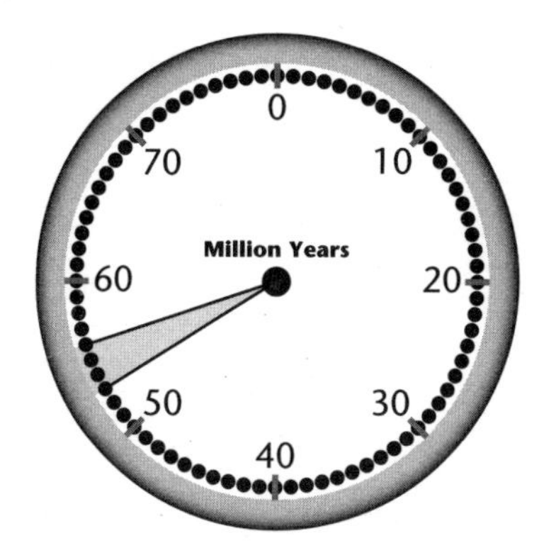

Imagine a stone that grows in the ground. A stone that is somehow alive. A stone without any outcrop, but which 'floats' in the soil. A stone that was once the remains of a human body, transformed. A stone that preserves the spirit of an ancestor.

Five thousand years ago, the sarsen stone could be all of these. How do we know? From how people employed and cherished sarsens, as well as what has come down in folklore.

A century ago, the sarsen's cousin, 'puddingstone', was known in Essex as the 'growing stone', believed to grow spontaneously in the soil and used as a lucky charm to protect against witchcraft. Revisiting the same boulder, people convinced themselves the stone had spawned (like the massive spherical sandstone concretions, known as 'trovants', found in the Romanian village of Costeşti). In a 1662 'witch' burial at Aldenham, a puddingstone boulder was laid on the coffin to prevent resurrection.

Many churches in north Essex have sarsens in their foundations. In the churchyard at Alphamstone there are eleven large sarsens: it seems a pagan stone circle was later transformed into a church. In the year 2000, excavations of a group of 3,500-year-old Bronze Age dwellings in the path of a Stansted airport runway revealed that the community had dug a pit into which they placed a large ceremonial sarsen stone.

Close to the great sarsen stone circle at Avebury, the 40-metre-high and 167-metre-diameter Silbury Hill is the largest of all European late Neolithic monuments, dating from around 2400 BCE. Sarsen stones were carefully buried in a ring, within the mound, as though this was some kind of memorial to the ancestors. Mayan pyramids are believed to emulate capricious Mexican volcanoes. What was Silbury Hill imitating?

*

Of course, the most celebrated use of sarsens was in the construction of Stonehenge (Fig. 25). Here the unique spirituality of the materials intersected with a daring architectural and astronomic vision. The design of the monument first required a deep appreciation of sarsen resources. The architect (or commissioning council?) must have walked among the scattered stones and commanded that the largest should be included in the construction. The order was for thirty sarsens that could be fashioned into blocks 4.1 metres high and 2.5 metres wide, each weighing 25 tons. Sheltered within the stone circle there were five even more massive stones, arranged in a horseshoe, each between 6 metres and 7.3 metres long. In the central trilithon, the stone listed as '56' is almost 9 metres long and weighs 50 tons. (Its partner upright has collapsed.)

We can guess the transportation team's initial reaction: denying vehemently such a stone could even be moved. But the architect(s) won the argument. The largest of all stones had to be incorporated in this prestigious and unique temple.

Figure 25. Sarsen henge and a pristine sarsen field at Fyfield Down

Unlike the Avebury stone circles, made from raw, unshaped sarsens and situated in the midst of the sarsen fields, the site at Stonehenge must have been especially potent. Irrespective of the labour and the cost, the sarsen stones would have to be transported 25 kilometres to this sacred location. Four thousand five hundred years later we can only guess the reason why.

The largest stones at Stonehenge exceed the dimensions of any remaining wild sarsens.[1] No stones longer than 5 metres were used in any other barrow or stone circle in the region. This scarcity elevated the status of the henge: it would never be possible to build a rival.

Fifty out of 52 sarsens at Stonehenge were found to have very similar surface geochemistry, suggesting they shared a common origin. More precise tests required samples from the stones, but these are now prohibited. Fortunately in 1958 three cores had been cut from a fallen Stonehenge sarsen, in order to fit a metal tie to prevent the stone splitting. Sixty years later rock from one of these cores was unexpectedly returned from the USA, by an engineer involved in the original coring who had kept it as a souvenir.

The geochemical 'fingerprint' of this core matched the surviving sarsens from a 600-hectare area of the West Woods on the Marlborough Downs, south of the River Kennet. West Woods appears to have been the source of almost all the Stonehenge sarsens.[2]

After the Neolithic transport team had made their latest delivery, the masons were tasked to batter the stones with sarsen and flint hammers to finish shaping their rectangular profiles, carefully carving the tenon joints into which the lintel mortices would fit.

The Stonehenge sarsen circle was 'married' to an inner circle of 'bluestones'. The precise source has now been matched to a specific rocky outcrop of metamorphosed Ordovician age dolerite at Carn Goedog, on the northern flank of the Preseli Hills, in Pembrokeshire, south-west Wales. In the 100-metre-wide slab pile, the fissiparous blocks appear as if suspended in some subterranean explosion. For reasons unknown, around 5,000 years ago, a set of up to eighty bluestones, initially part of a local stone circle, each weighing 2–4 tons, were transported overland, perhaps part way by boat, 200 kilometres east to Salisbury Plain. Forty-three survive.

Were these trophies of war or some symbolic merger of cultures? Maybe there was a practical reason? For the late Neolithic stone circle designer, there was no technology by which to cut slabs from hard bedrock. Sarsen and bluestone, yin and yang, are united by one key property. Both arrived massively 'pre-sliced'.

The Preseli mountains escaped the glaciers through the Ice Ages: the only upland location in Wales to remain ice-free.[3] If the 'exploding' rock pile at Carn Goedog had been overtaken by ice, the rocks would have been scattered for miles and buried in boulder clay. Instead the Ice Age freeze-thaw split the rock into slabs 2 or 3 metres long and one-third of a metre thick and then left them in an accessible heap. Nowhere else in Wales offers such ready supplies of tough pre-shaped slabs, perfect for a local stone circle, and then evidently worth transporting all the way to Salisbury Plain.

The 6-ton Altar Stone at Stonehenge was different again: a purplish-green micaceous sandstone, possibly from south Wales. We can assume each provenance brought symbolism, perhaps the uniting of faiths.

Although sarsens are found south of a line from Dorset through to Norfolk, the mother lode is in central Wiltshire, where the landscape was once cluttered with the great stones. At a few locations the unploughable, pristine sarsen fields were never cleared. At Fyfield Down (Fig. 25), the greatest concentration of natural sarsens *in situ*, the stones resemble an endangered herd, sitting quietly, the last remnants of the great sarsen-filled prairie that once extended for miles. Before the prairie there was sarsen-filled woodland (as still surrounds Ashdown House in Oxfordshire). The majority of the surviving stones are babies, three-quarters of them less than 1.5 metres long.

With no bedrock, once the largest slabs were removed the resource was gone. These strange 'floating' tabular boulders have provoked much speculation. Christopher Wren thought they had been blasted out of a volcano. William Stukeley claimed they had some anti-gravity property whereby the Earth's rotation raised them out of the ground.

Daines Barrington believed a great earthquake had scattered them across the landscape.[4] Eventually geologist George Greenough identified that sarsens were formerly a layer of rock that had overlain the chalk.

★

Very rarely, sarsen stones have been found geologically *in situ*, located in the mixed shoreline sands and gravels of the end-Paleocene 'Reading Beds'. Sarsens started their life as beach sands along the banks of rivers and the sea, reflecting a marine high water when the North Sea was isolated from the world ocean. Why did this single seam of sand become alchemically transformed into quartzite sarsen? Many other layers of sand, laid down through the Paleocene across south-east Britain, have survived in the loose state in which they were deposited.

To get silica to dissolve in water you need two ingredients: high temperatures and acidity. Some 55.5 million years ago, there was an extraordinary episode of global warming known as the Paleocene/Eocene thermal maximum, caused by a massive injection of carbon dioxide and methane into the atmosphere (and detected in carbon isotope concentrations). Although this lasted less than 20,000 years, global temperatures increased by 5–8°C and warming was then sustained for another 200,000 years. The effects were strongest around the north and central Atlantic, most likely from basaltic volcanoes and dyke swarms rupturing oil and gas reservoirs as the new North Atlantic spreading ridge got going.

There is still debate whether silicification occurred at the surface or a few metres underground. Beyond the high temperatures the additional ingredient was water acidified by contact with pyrite (iron sulphide) and enriched with aluminium from the breakdown of clays and feldspar in volcanic ash.[5] As the climate warmed, water was evaporating from the surface of the sand, drawing more acidic hot water behind it, dissolving and recrystallizing the silica. The separate grains of sand became fused. After a few thousand years the sand had transformed into a thick layer of 'silcrete', in effect a '100% silica' concrete.

Later the layer broke into slabs, foundering when the underlying sands and clays were eroded away.

Sarsens are a memorial to the strongest episode of global warming in the last 66 million years—an episode which, thanks to our reckless burning of fossil fuels, we have every chance of exceeding. Many species of animals became extinct at this time, unable to cope with the impact of dramatic warming on ecosystems.

Sacred to Neolithic man, sarsens present a potent warning for our future.

★

In Hertfordshire and parts of Buckinghamshire, in place of silicified sands we have a rock composed of rounded flint pebbles, cemented with silica, known for its plummy appearance as 'puddingstone'. All the puddingstone in Essex has come from the west, picked up in fierce glacial floods of the former River Thames and transported from Hertfordshire.

The flint pebbles in puddingstone can be up to 5 centimetres in diameter. The rock has been marinated in silica so thoroughly that it does not even fracture around the pebbles and can be used for jewellery.

Puddingstone is contemporary with sarsen, formed in the same episode of global warming. In place of beach sands that solidified into sarsen,[6] gravel shingle was converted into puddingstone. At the beginning of the Paleocene, the strongest land uplift ran NW–SE across central England, and more than 500 metres of chalk was eroded, leaving a residue of flint pebbles, just as today the shingle beaches at Dungeness and Chesil Beach are the residue from the erosion of billions of tons of raised and eroded chalk rock.

When found *in situ*, the puddingstone pebble bed is between 20 and 90 centimetres thick. Near St Albans, a quarry exposes the original layer of Hertfordshire Puddingstone which can be traced for 10 kilometres.

At the Colliers End and Puckeridge sites the seam was avidly mined both in Neolithic times and by the Romans. They pulled out massive slabs of puddingstone from which to manufacture 'querns' for grinding grains of wheat[7] (Fig. 26). No pottery could survive such battering, so the quern was patiently carved from a single unfractured block,

Figure 26. Roman puddingstone quern

comprising a solid stone pestle with a matching stone mortar. Silica puddingstone was so hard it was very difficult to shape, but that was the point. An inferior sandstone quern adulterated the flour with teeth-wrecking grains of sand. The silica surface was rough, scratching open the wheat grains.

For a century between the Roman invasion and 150 CE, across eastern and southern England, puddingstone querns were the fashion item in any well-appointed Anglo-Roman kitchen. Some 700 have been found, each pestle and mortar pair carved from an 80-kilogram block of puddingstone.[8]

Yet puddingstone was difficult and, from the sharp rock splinters, even dangerous to carve. After 150 CE inferior, but cheaper, vesicular lava querns were being imported from the Eifel volcanoes close to Bonn.

So whereas the late Neolithic builders of Stonehenge stripped the landscape of great sarsen slabs, it seems it was the Romans who depleted Hertfordshire of its puddingstones.

★

At the Ona lighthouse on the west coast of Scotland, on the evening of Easter Monday, 29 March 1875, the keeper noted in his log, 'a fine grayish sand fell with the rain, forming a layer two lines thick which stuck to window panes and house walls'.[9] The same evening farmers from Trysil, close to the Swedish border with Norway, noticed their eyes stinging from falling dust. Later that night in Stockholm two walkers found their clothes covered with yellowish-grey dust, which also dimmed the street lamps.

It took a few days to register; north-westerly winds had brought ash from the eruption of Askja Volcano in Iceland.

Around 56 Ma, sea-floor spreading between Greenland and Europe began with a furious episode of regional volcanic activity. On the Faroe Islands, 6,600 metres of lavas accumulated: more than three times the thickness laid down in the Hebrides a few million years earlier.

Iceland is just the latest manifestation of spreading ridge volcanoes above the waves. At the origin of the north Atlantic Ocean, there were the equivalent of, perhaps, ten volcanic 'Icelands' all along the new plate boundary, extending in a wide zone of rifting and lava fields from west of the Shetlands and down the whole length of east Greenland (Fig. 27), with scores of erupting volcanoes sending thick clouds of ash downwind.[10] And it was not only ash.

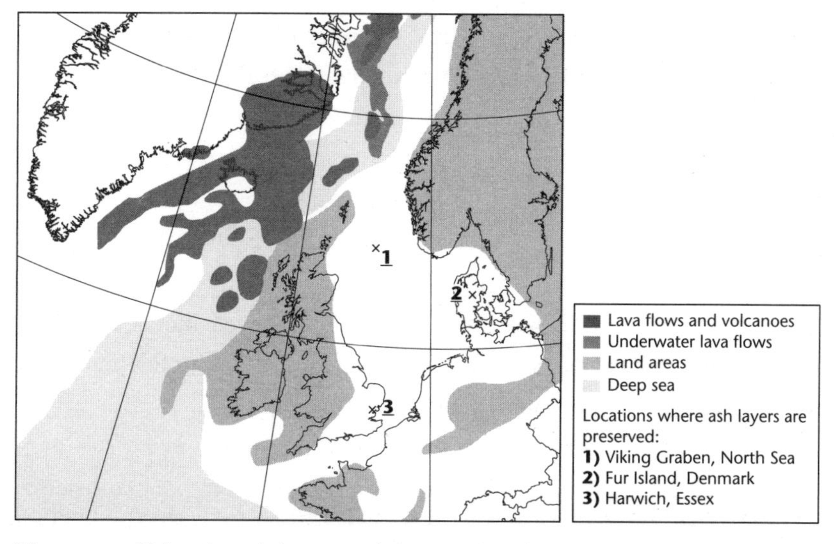

Figure 27. Volcanic activity around the North Atlantic in the Early Eocene

In the 1783 Laki eruption, the haze made the Sun appear red as blood. People who 'suffer[ed] from chest ailments could hardly breathe and nearly lost consciousness'. The acid rain caused timber to lose its colour while the grass, 'began to fade and wilt'. Half the cattle and three-quarters of all the sheep and horses in Iceland died of starvation over the following winter and spring, causing a famine that led to a dramatic rise in human death rates. The toxic haze spread to Europe, where it caused widespread health problems and mortality.

For Britain, the beginning of the Eocene was the most polluted period in its history, only rivalled by the burning of household and industrial coal in urban centres through the nineteenth and early twentieth centuries.

The largest eruptions left their ash-layer 'calling cards' in the sedimentary record, far downwind from the volcanoes, in and around the North Sea.

In a former lake on the island of Fur in northern Jutland, free from other local sources of sediment, there are more than 200 ash layers from this period, most of them basaltic (Fig. 28). (During the Ice Age the layers were bulldozed into cliffs by the advancing Scandinavian ice sheet.)

The layers record two episodes. The older comprises a small number of ash eruptions from the British Paleocene volcanoes.

Above these layers there is a much more prolific set of eruptions launched from the new spreading ridge volcanoes farther to the west. Most ash layers are only the width of window glass but some are book-thick. Alongside the ash layers there are beautifully preserved fossil fish, reptiles, insects, birds, and plants.

The same set of ash layers are found in borehole cores from the northern end of the North Sea. Again the record starts with the sporadic eruptions of the Hebridean volcanoes, followed by the stripes from hundreds of ash layers. Some of these eruptions were enormous.[11] The 20-centimetre-thick 'Layer 13' at Fur has reduced to 2 centimetres in a fossil lake bed in Austria. Imagine the blighted trees and choked animals from 20 centimetres of ashfall across the whole of Britain.

Figure 28. Early Eocene volcanic ashfall layers (brought vertical by a much more recent ice sheet), on Fur Island, northern Jutland

As seen in cliffs at Wrabness in Essex, the base of the London Clay contains thirty of the younger ash layers, each 10–80 millimetres thick. Higher up the sequence the individual ashfall layers become blurred. The London Basin was not only receiving ash from the sky, but ash from the same eruptions slowly washed out of forests, into streams and rivers draining to the south-east.

Add together the thickness of the individual ash layers cored in the northern North Sea and they sum to 8 metres. Perform the same exercise at Fur in Denmark, farther from the volcanoes, and this reduces to 4 metres of ash. If we assume an average of 6 metres of ash fell across the landmass of Britain, that would equal 600 cubic kilometres, much of which then drained in rivers to the south-east. This volcanic ash is a primary constituent of the London Clay formation that underlies most of London north of the River Thames.[12] The consistent composition indicates that, for at least a million years, there was a continuous supply of volcanic ash material, carried on north-westerly winds and transported in swollen, muddy rivers, from the distant spreading ridge volcanoes.

Fragments of volcanic glass in the ash have decomposed into a clay almost impermeable to water, making it the perfect medium for tunnelling and deep foundations. Seven hundred kilometres of tube tunnels have been cut in London Clay, giving the city a thirty-five-year advantage in underground transportation over Berlin (underlain by glacial sands) or New York (founded on metamorphic schist).

In Essex the London Clay is 150 metres thick, tapering to the west from later uplift and erosion. The clay was deposited in a shallow sea, adjacent to a swampy mangrove, Sumatra-style, delta coast, with luxuriant tree ferns and palms tilting over the beaches. West towards Reading was sandy while the east was glutinous clays, decomposed volcanic ash, perfect for sealing and preserving leaves, fruits, shells, and bones. Among the wealth of fossils: the remains of more than 350 species of plant. Turtles, crocodiles, and snakes lived alongside a range of mammals and birds including owls and parrots. In the sea swam sharks and rays, feasting on the floating nautilus cephalopod, while on the sea floor scuttled crabs, lobsters, and shellfish. London has never seen such biological richness.

Before the arrival of the railway, ordinary London buildings were manufactured out of London Clay.

The London stock brick is not a thing of beauty—it is crude, often misshapen, pock-marked, and mottled with scabs and cracks and harsh colours from pale yellow to grey, purple patches, reds and oranges, and then black where attacked by the smog. Every brick is unique and handmade. As the developers commissioned new streets in Georgian London, from Hackney to Notting Hill, at the edge of each development they dug down into the London Clay and, using a mould, cut rough artisanal bricks that were stacked alongside wood scraps and waste to be fired in a brick kiln. The smoke and scars created by this process were almost as ugly as the conditions under which the ash had rained down from the sky, obliterating plants and fouling streams. The well-off ladies and gentlemen of London must have looked on in horror as the green fields were scarred with excavations out of which were built the endless row houses. London Clay was useless for agriculture (termed 'bringing up poison' on account of its impermeability), but at the same time it was endlessly fertile for breeding houses. In a single summer, a gang of six brickmakers could manufacture a million bricks—enough for thirty homes. As the buildings were completed, the gang shifted their diggings to the neighbouring street.[13]

In a cartoon from 1829 ('London Going out of Town or the March of Bricks and Mortar'), George Cruikshank illustrates the furious digging, the polluted skies, and a kiln spewing bricks, like an erupting volcano, to overwhelm the rustic landscape of trees, meadows, and haystacks.

This activity only came to an end in the 1840s with the arrival of the railways, when great brickwork quarries opened in the Midlands to manufacture a much more uniform, but much less interesting product, machine-cut from older seams of clay, generally in a monotonous 'brick-red' colour.

Thereby the original buildings of London (including the house in which this chapter is being written) were constructed out of new oceanic crust erupted by volcanoes 1,000 miles away.

Today London Clay brings blight not only to farmers, but to all those whose houses rest on clay, which shrinks when thirsty tree roots suck out the residual water, swelling again through a wet winter. Shallow-foundation brick buildings grow warped and torn in the months after a summer drought. The North Atlantic volcanoes both made and undermined the buildings of London.

London was originally built on, and built out of, the products of a spreading ridge plate boundary. London, one could say, 'is a plate boundary volcanic city'.

★

We are now 10 million years into the Cenozoic. From 60 down to 55 Ma has been a period of frenzied activity. A vast mantle plume raised land from out of the chalk sea. Two attempts have failed to tear the land apart and carve new plate boundaries right through Britain. A new spreading ridge has opened on the far side of the Rockall Bank and to the west of the Shetlands. The ash eruptions and gigantic hydrocarbon fires have been polluting Europe with toxic fumes, but finally the pollution is clearing as the volcanoes are slowly sliding underwater.

And yet the tectonic situation around Britain is far from calm. Faults are starting to move, plate boundaries to reconfigure.

To tell this tectonic story we should recall the trinity of fault styles: 'normal' (also known as extensional), 'reverse' (also known as thrust), and 'strike-slip'. We can summon all that we learnt in that contemplation of the crack in the wall.

With plentiful magma to grout up the gaps, horizontal stretching leads to dykes. Without the magma we get extensional normal faults.

We will be switching perspectives between the grand architecture of the plates and the local ornament of displaced strata. There is an underlying logic to tectonics that brings order and consistency to what might seem to be disconnected observations. Yet tectonics has some mystique of difficulty, linking terminology to spatial conceptualization that may have discouraged its inclusion in the corpus of popularized science. We will follow the local tectonic paths, like narrative threads, as they become woven into big-picture stories. As the captain says, 'The ride is about to become a little bumpy. Please hold on tight.'

6

The foothills of the Pyrenees

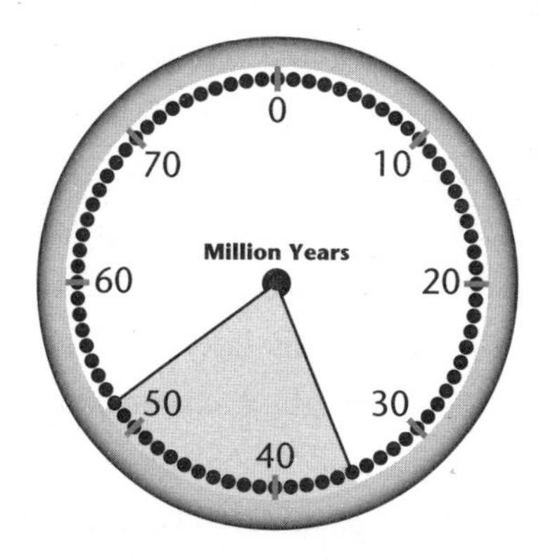

For a few months before going to university, I got a base-level intern job working for the Government Survey, tasked to map the offshore geology of the UK continental shelf.

For three weeks on a survey ship criss-crossing the Moray Firth off the coast of north-east Scotland, we collected the geophysical data (seismic reflection, side-scan sonar, gravity meter, and magnetometer).

The ship towed a frame with floats carrying a '1,000 joule sparker array', which every few seconds emitted an electrical spark explosion in the water. Trailing further behind was a hydrophone, to pick up the echoes from the seabed and from layers within the sediments beneath. The echoes were printed directly on a paper roll, one rusty line for each explosion, a few seconds apart, so that the paper profile captured the consistent pattern of echoes as the ship maintained its course. The surveying continued day and night, while we kept watch to check the equipment didn't malfunction, looking out for the radar reflection of a trawler laying its nets, and glancing at the paper profile to catch a glimpse of the hidden sea floor, 80 metres below our keel.

These seismic reflection surveys revealed some of the invisible geology beneath the sea: buried landscapes, traces of layers, folds, and faults. (They were called 'seismic' because they used the properties of sound waves—like 'seismic' earthquake vibrations—passing through the water column and reverberating off the sea floor and strata beneath.)

The primitive equipment we were using was puny compared to the capabilities now operated on behalf of the big oil companies, with their powerful airguns, multiple hydrophone streamers, and enormous data downloads. Back in the computer centre this mass of data would be analysed with the latest 3-D mapping software, to visualize a detailed picture of the layers of rocks and faults beneath the seabed, hundreds of metres underground. It is these techniques that have enabled the geology and subsurface to be mapped all around the British continental shelf.

Collecting seismic profiles is much cheaper offshore, because the whole array of seismic sources and hydrophones can simply be towed behind a ship, rather than having to be laboriously planted and removed from the ground. (On land, the menu of vibration sources becomes 'à la carte'. Beyond enthusiastic sticks of dynamite, geophysicists have employed road compactors, lead weights dropped from a height, sledgehammers swung onto a metal plate, and even a machine gun, firing slugs into the ground.)

Sixty years ago, we knew nothing about the geology under the seas around Britain. Today we have intricate three-dimensional imaging that reveals far more tectonic information on when each fault moved than is available from mapping the rock outcrop on land.

*

Geologists use the term 'basin' where a thick sequence of strata has accumulated. Basins come in all sizes: from a 'pull-apart' at the jog in a strike-slip fault, a few kilometres across, to the scale of the whole North Sea.

To produce a basin takes 'normal' extensional faults, which form when the crust is stretched. Layers of sand and mud accumulate on the downfaulted 'hanging wall' of the fault. If the faults face one another, you have a rift valley. Or the faults may all dip the same way, like a set of slumped books on a shelf.

Typically, even after the normal faults have stopped extending for tens of millions of years, the basin continues to sink. When the upper crust is rifted apart, the hot lower crust and mantle are stretched and thinned like chewing gum. This draws up deeper, hotter mantle material, which is why volcanic activity often accompanies rifting. Once the rifting has stopped, the raised temperatures beneath the crust begin to chill as the heat drains out at the surface. The heavier mantle rock contracts and weighs down the crust so that even more mud and sand accumulate across an even wider basin. The load of sediment itself also makes the crust sag further. Geologists, under a Texan influence, call the cross section through a 'mature' basin a 'steer's head profile', with the thicker, older strata concentrated up against the deeply buried faults of the rift, whereas the youngest layers span the whole basin (Fig. 29).

In a sinking basin, as the layers become more deeply buried their temperature rises, cooking any organic material and releasing oil and

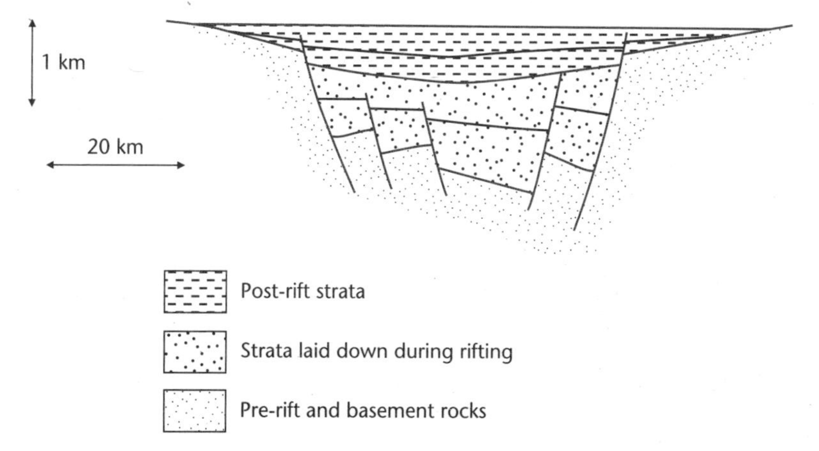

Figure 29. 'Steer's head profile' across a mature rifted sedimentary basin

gas. The hydrocarbons may leak up through the strata, along fractures around faults, for example, all the way to the surface. Sometimes this flow is blocked by an impervious geological layer. If this seal is domed, it can trap rising oil or gas, like an upside-down version of a land-based water reservoir.

From 200 to 100 Ma the crust around Britain was being stretched and opened across a network of rifted basins. The great supercontinent Pangaea was starting to pull apart. A prominent set of E–W trending, mostly southerly dipping, extensional faults formed across southern England (Fig. 30). To the east through Hampshire and Sussex the Weald Basin was sliced by a swarm of smaller faults. Farther to the west lay the Wessex Basin carved by a few big faults and to the south the Channel Basin, with two mega faults, one of which follows the east–west axis of the Isle of Wight.

The Weald Basin today is not just a rifted basin, but also a great hollowed-out, raised-arch 'anticline'. Anticline means 'opposite slope', like a pitched roof or an arch. The anticline was formed by horizontal

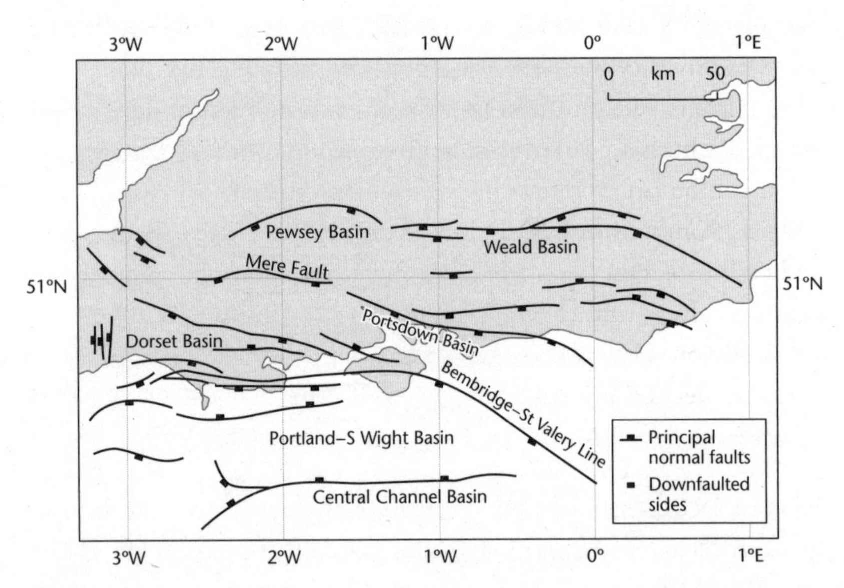

Figure 30. Extensional Jurassic and Early Cretaceous faults beneath southern England

north–south crustal shortening, long after the original rifting and extension. In plan-view the Weald anticline is shaped like a lozenge with older formations exposed towards the centre, extending from Hampshire through Sussex and Kent into the French coastal 'Boulonnais' south of Calais. As the anticline arched and raised, it became eroded. Some 1,500 metres of strata have been lost from the eastern core, although the hills probably never rose higher than 500 metres.[1]

The Weald is also an 'anticlinorium', filled by a crowd of lesser 'baby' anticlines.[2] The baby anticlines share many features in common. They all run more or less east–west. They tend to lie 'en echelon', like rucks in a rug on a polished tile floor. When one anticline dies out to the east, or west, another nearby anticline rises to take up the slack, partly overlapping but not quite on the same alignment.

Some of the Weald anticlines can be traced into the Wessex Basin to the west, where the anticlines are larger and more widely spaced.

Calling them all 'anticlines' is not quite correct—they are really 'monoclines' (meaning 'one-sided slope'). South of the highest point of each fold, the layers are nearly horizontal. Layers on the northern side of each Weald monocline typically 'dip', or slope 15–25 degrees or more down to the north. The slopes are even steeper along some of the larger faults in the Wessex Basin and in the Channel. On the Isle of Wight, as we have seen, the strata located on the northern side of one of the largest monoclines are almost vertical.

Apart from in the cliffs on the Isle of Wight it is not easy to go out and see one of these 'monoclines'. Some were exposed in a nineteenth-century railway cutting or village stone quarry and are now shrouded in vegetation. Others have only been seen in subsurface mapping of seismic reflection profiles.

★

I received a call from seismologist Geof King on the day after the biggest earthquake in Algeria for a generation. I had come to know him on a Royal Geographical Society expedition in that summer of 1980 when he was running a network of seismic recorders in the Karakoram

Mountains and I was writing features on mountain science for *New Scientist*. He was going to drive out to Algeria in a Geophysics Department Citroën and was looking for a co-driver for the Toyota, which was loaded with seismic recorders. At the end of a research fellowship and jobless, I was intrigued by what could be learnt from a big earthquake.

Rainy autumn was well under way in Cambridge, but it was still warm in Marseilles where we got on the ferry to Algiers. On arrival the jeep was impounded because we didn't have the correct import papers. With my Indian seismologist co-driver Harsh Sinvhal, we spent a night in a flea-ridden hostel in the casbah. The following day, after many calls and through the intervention of the Minister of Construction, we cleared customs and late the next day we were amidst the shattered villages around the ruined city of El Asnam, beginning to piece together the disaster. What could be learnt that would help future cities identify where to expect such dangerous earthquakes?

The magnitude 7.3 shock was caused by 3 metres of upwards displacement on a reverse fault located beneath the city of El Asnam.[3] Before this earthquake, geological wisdom taught that folds formed gradually as a result of slowly accumulating forces. Only faults broke suddenly.

Part of the fault rupture at El Asnam had reached all the way to the surface at the base of a range of hills, emerging beneath a railway track on which a train had the misfortune to be passing. Yet for most of its length the fault stayed hidden underground.

The Cheliff River cut through these hills. After the earthquake the river stopped flowing. A wide lake began to fill in the plain upstream.

New lakes are common after big earthquakes, but there was no obstructing landslide, nor had the hidden fault dammed the river. Instead, a few kilometres downstream, culminating at the centre of the line of hills, the riverbed had risen. As mapped by the expert on Mediterranean valleys Claudio Vita Finzi, a prominent river terrace, marking a former flood plain, also reached its highest elevation where

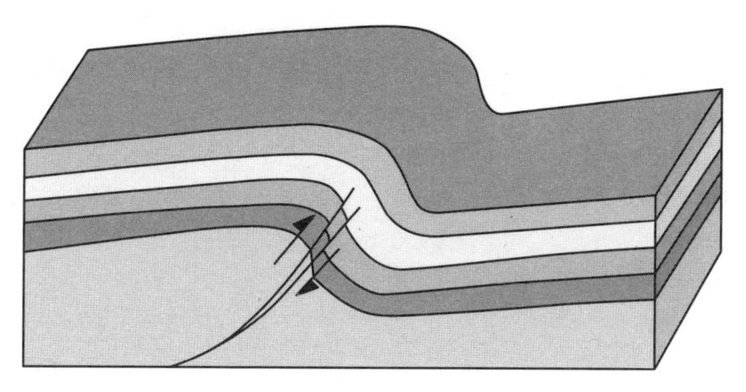

Figure 31. Blind thrust and overlying monocline

the river was backed up. The anticlinal arch had become more tightly folded in the latest earthquake. The river terrace showed the latest arching had repeated the pattern of folding in previous earthquakes.

The earthquake had revealed some new science.[4] It was not that a fold happened slowly and a fault moved fast. The sudden buried 'reverse' fault movement had tightened the overlying fold within the tens of seconds it took for the fault to rupture (see Fig. 31).

Over the next few years these lessons were to be repeated after several earthquakes in California.[5] The first, in 1983, occurred at the Spanish-sounding town of Coalinga (which turned out to be the prosaic 'Coaling A'—a refuelling stop on the transcontinental railway). A reverse or 'thrust' fault that does not break through to the surface is termed 'blind'. Typically, with sudden movement on a blind thrust fault, the overlying fold becomes more tightly warped. All around Los Angeles there are folds, capping active blind thrust faults. The longer the mapped length of the structure, the more likely that the underlying reverse fault breaks all the way to the surface, or 'daylights'. However, earthquakes on blind thrust faults could be every bit as dangerous as where the underlying reverse faults cut through to the surface.

★

Based on what was first learnt on that scientific research mission in Algeria back in 1980, we now understand that each Weald monocline is

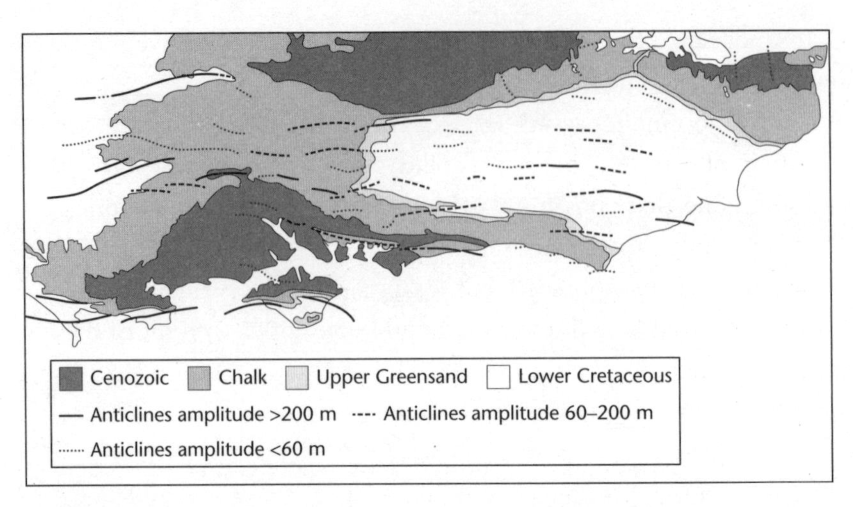

Figure 32. Anticlines (monoclines) in and around the Weald Basin

underlain by a blind (or 'buried') southerly dipping thrust fault (Fig. 32). And these thrust faults have simply reversed the movement on previous extensional faults.

Southern England provides a rich hunting ground for both blind (hidden) thrust faults, with their overlying monocline folds, and those in which the underlying reverse fault has broken through to the surface.

A series of 'en echelon' monoclines run through mid-Surrey. For the Romans the 12-kilometre, arrow-straight, 'Hog's Back' monocline ridge was an embankment crying out for a road, and today the A31 highway from London to Winchester has simply tarmacked the Roman cobbles. In places, the underlying fault cuts through the steep fold, exposing up to 120 metres of reverse displacement,[6] but mostly the fault stays hidden.

The monoclines are larger and more widely separated to the west in the Wessex Basin. The eastern end of the Vale of Wardour Anticline is a fault with 370 metres of displacement while the folded sediments dip at 40 degrees down to the north. Something to contemplate while stuck in a traffic jam on the A303, which follows the valley eroded where the underlying Mere Fault[7] thrusts weak Kimmeridge Clay over younger, and tougher, chalk.

To understand the structural architecture of this region we need to revisit some tectonic history.

Around 300 million years ago, during the formation of the super-continent of Pangaea, continents collided along an east–west 'Variscan' collision zone that runs from present-day Belgium, under southern England through the Bristol Channel, and into southern Ireland. The crust absorbed the collision by stacking a series of shallow dipping thrust faults, to raise a line of mountains. The northern limit of these thrusts is called the 'Variscan Front'. The mountains got eroded away, whereas the underlying fault architecture remained.

These 'Variscan' thrust (reverse) faults, dipping at 20–30 degrees to the south, cut the Somerset coalfield—as encountered by William Smith in his first mine surveys.[8] Almost identical faults cut the strata in coal mines in western Belgium. Much the same thrust faults must lie deep beneath the Weald Basin. Unlike the situation in Somerset or Belgium, between 100 and 200 Ma the crust beneath the Weald became stretched north–south, and the thrusts reversed their original displacement to become the roots to extensional faults (Fig. 33). Across the Weald Basin the crust stretched by 15 per cent.

Drilling for oil on the Isle of Wight structure revealed,[9] to some surprise, that even though it was the southern side of the fault that has latterly risen by hundreds of metres, the layers of Cretaceous and Jurassic sediments are much thicker south of the fault than to the north. Like the structures in the Weald, the fault had seen more than a kilometre of extension, before it was reactivated with about half that amount of reverse displacement.

Since the start of the Cenozoic, the underlying southerly dipping faults have gone back to their original function, becoming reverse faults absorbing some of the subsequent crustal shortening. However, this reverse fault displacement has been less than the previous extension, which was less than the original Variscan thrusting. Push, pull, push. Think of this as 'accordion' tectonics, or maybe like the movement of a muscle, stretching and contracting.

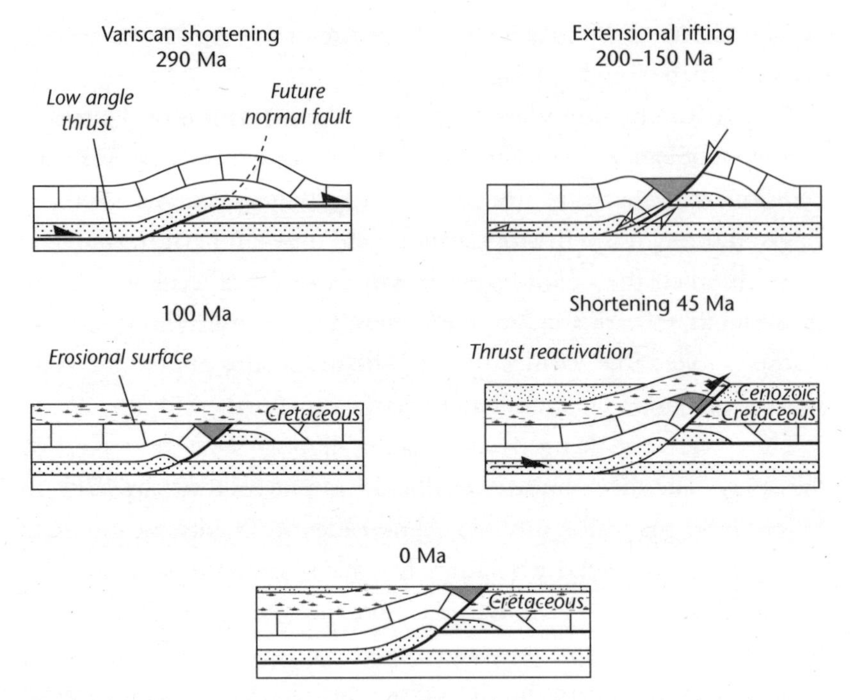

Figure 33. Basement fault reactivation

The Isle of Wight structure is by far the largest of the onshore (semi-)blind thrust faults with their overlying monoclines.[10] To the west, after 28 kilometres this structure returns on-land at another set of contorted chalk rock stacks at Ballard Down in Purbeck. The fault then runs west along the Jurassic Coast to form the outer wings of Lulworth Cove.

Tectonics has created a theatre: the beach is the circle, the sea the watery stalls, stirred by refracted curving waves. In *Far from the Madding Crowd* Thomas Hardy gets one of Bathsheba's suitors, Sergeant Troy, to plunge into the cove, swimming out 'between the two projecting spurs of rock which formed the pillars of Hercules to this miniature Mediterranean'.

Two kilometres farther west, the northern dip of the Portland Beds reaches 80 degrees in the Durdle Door sea arch (Fig. 34) with 750 metres of uplift on the underlying monocline. The Jurassic sediments

Figure 34. Reversed fault and overlying fold exposed at Durdle Door, Jurassic Coast

comprise a mix of hard and soft layers which have been crumpled like a dropped layer cake in the heart of the fold.

Stratigraphically this may be the 'Jurassic Coast' but it would be nothing without the Cenozoic tectonic architecture. Beyond Durdle Door the outcrop of the hardest Purbeck stone has left a line of sea stacks and reefs known by fishermen as The Bull, The Blind Cow, and The Cow and Calf. The monocline continues resolutely westward, just offshore from Bat's Head and White Nothe where the chalk is steeply inclined and in places overturned.

The westward path of the reversed fault and monoclinal fold, which began in the Isle of Wight, ends out of sight beneath Weymouth Bay. Part of the shortening (crack-in-the-wall style) switches into the NNE-trending 'Mangerton Fault', with a kilometre of horizontal sinistral displacement, which comes onshore through Bridport. Inland, the tectonic complexity, juxtaposing kilometre-scale blocks of resistant Jurassic sandstone and butter-soft erodible clay sequences, has created the most beautiful Dorset landscapes. To the west of Beaminster and around Symondsbury, the steep, smooth pasture knolls and ridges seem carved by a geomorphic Henry Moore, then garnished with summit groves and hanger woods.

To the east of the Isle of Wight, the structure swings towards the south-east as the Bembridge–St Valery Line and continues through northern France as the Pays de Bray Fault system. (Like 'the crack in the wall', the offset looks different where the fault changes orientation.) The Pays de Bray Fault has principally carried up to a kilometre of dextral horizontal strike-slip movement. Traced to the south-east, after forming the harbour at Dieppe the fault has generated an overlying fold between Neufchâtel-en-Bray and Beauvais, beyond which it brings some welcome Eurostar-window scenery where intersected by the TGV line from Lille to Paris.

South of the Isle of Wight, in the middle of the Channel (Fig. 30), there is a similar monoclinal E–W structure, feeding off the Pays de Bray Fault, but the mirror image of the Isle of Wight structure with extension down to the north and the subsequent climax of the later reverse fault uplift also on the northern side.[11] Not so long ago there would have been a sweeping mid-Channel chalk hilltop ridge along the line of this structure, now eroded away by the sea.

To date when the Isle of Wight and Weald reverse faults were moving we need geological clues. Near Aldershot in Surrey, stones of Lower Cretaceous 'Wealden' (120 Ma) age turn up in a pebble bed[12] deposited 42 Ma ago. Erosion had penetrated through the thick roof of chalk that formerly covered the older Weald strata.

More precise dates come from deposits eroded off a neighbouring active fault. These recycled materials reveal two phases of Eocene tectonics: one in the period from 41 to 48 Ma and the second in the period from 38 to 41 Ma.[13]

Massive beach-rounded flints, weighing up to 10 kilograms, turn up in the 42 Ma Boscombe Sand in Dorset, hinting that the chalk monocline between Purbeck and the Isle of Wight had already been breached. At Creechbarrow Down in Dorset, gravels, 100 metres thick, are full of huge flints sourced from the neighbouring fold, contemporary with the second phase of movement, dated to 40 Ma.[14]

★

To make sense of what was happening we need to look south. After 10 million years during which the African landmass ceased pressing on Europe, collision was resumed around 52 Ma. Present-day Spain and France were then on different tectonic plates: Spain was in lock-step with Africa whereas France was part of the Eurasian plate. The boundary between them ran WNW along the line of the Pyrenees (Fig. 35). During the Eocene, the collision of the African and Eurasian plates drove mountain building along a 1,000-kilometre-long 'Pyrenean' front that extended farther both to the east and to the west than what is today the residual length of the Pyrenees mountains.

This was a full-on confrontation. The thick continental crust of Spain was forced down underneath the continental crust of France in a titanic collision. Great slices of the crust were torn off and thrust back to the south. The 'suture zone', the North Pyrenean Fault, which marks the original boundary of the plates, lies on the French side of

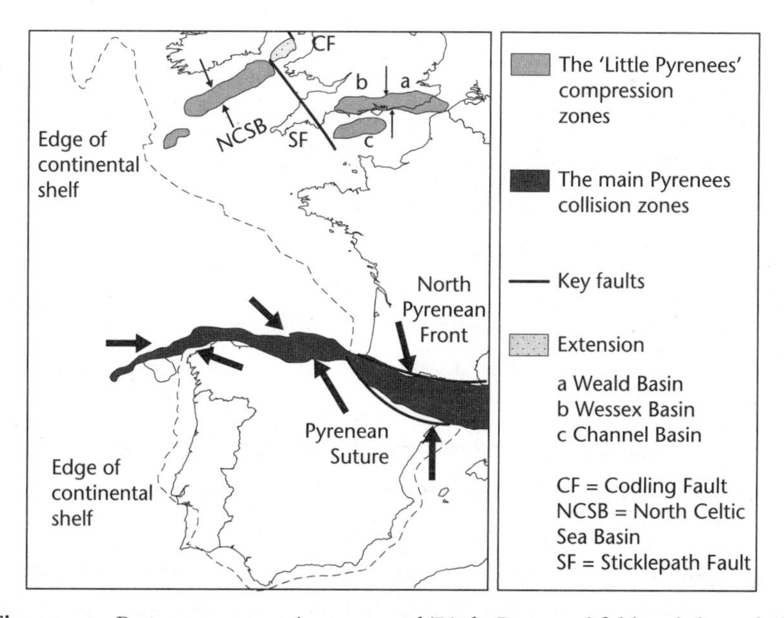

Figure 35. Pyrenean mountain range and 'Little Pyrenees' fold and thrust belt across southern England and south of Ireland in the Eocene

the mountains. The crumple zone across the mountain range is estimated to have absorbed 100–150 kilometres of shortening over a width of less than 150 kilometres.

Forty million years ago, along the northern coast of Spain, the Bay of Biscay ocean crust started to underthrust (to 'subduct'), to the south beneath the north Spanish margin.[15] Tens of kilometres of ocean crust were swallowed up. This subduction zone lies at the base of the steep and rugged continental slope of northern Spain, where today the sea floor descends to 4,500 metres in less than 50 kilometres.[16] The subduction zone originally extended to the west as far as the northern edge of the Galicia Bank off the north-west corner of Spain.

Eventually the Pyrenean collision zone choked as Himalayan-style mountains, perhaps 6,000 metres high, pushed back against the converging plates.

We can date the principal episodes of Pyrenean continental collision from the sediments laid down as the mountains were rising. On the French side there is a massive 'puddingstone' conglomerate known as the 'Poudingues de Palassou' full of boulders eroded off the collapsing mountainsides. This formation reveals the Pyrenean collision was at its height between 48 and 41 Ma. This conglomerate is overlain by undeformed late-Eocene strata. On the south side of the mountains there are folded conglomerates dating from a little later: between 41 and 37 Ma.

The Isle of Wight Fault was 'plugged' into France through the NW–SE Pays de Bray Fault. Fossils deposited at the south-east end of the Pays de Bray fold date when the fault was moving:[17] through the Eocene and just into the Oligocene (35 Ma). We believe the big Isle of Wight Fault moved in lockstep with the crowd of local faults that comprise the Weald Basin. Calcite veins, contemporary with the faulting in southern England, contain traces of uranium, which can be dated. Tectonics danced to a Pyrenean score, culminating in the late Eocene.[18]

In the era before plate tectonics, geologists dismissed the Isle of Wight fold as 'ripples from an Alpine storm'. More like big breaking waves than ripples, the Weald and Isle of Wight structures are really 'foothills of the Pyrenees', or perhaps 'the occasional frontier between the African and Eurasian plates'. Huge compressional stresses were leaking out to the north of the Pyrenees. Backed to the north by the stable and immobile massif known as the London or Anglo-Brabant platform, the underlying Wealden reverse faults were intermittently moving to pile up the overlying monoclines.

The pattern of reverse faults underlying anticlines can be traced far out to the west into the Atlantic. On the northern side of the Bay of Biscay seismic profiles reveal reverse faulting,[19] dated to 45–40 Ma, driving a swath of folding up to 50 kilometres wide, involving more than a kilometre of horizontal shortening.[20] When seen on seismic profiles these deep-water structures are particularly beautiful (Fig. 36)—huge

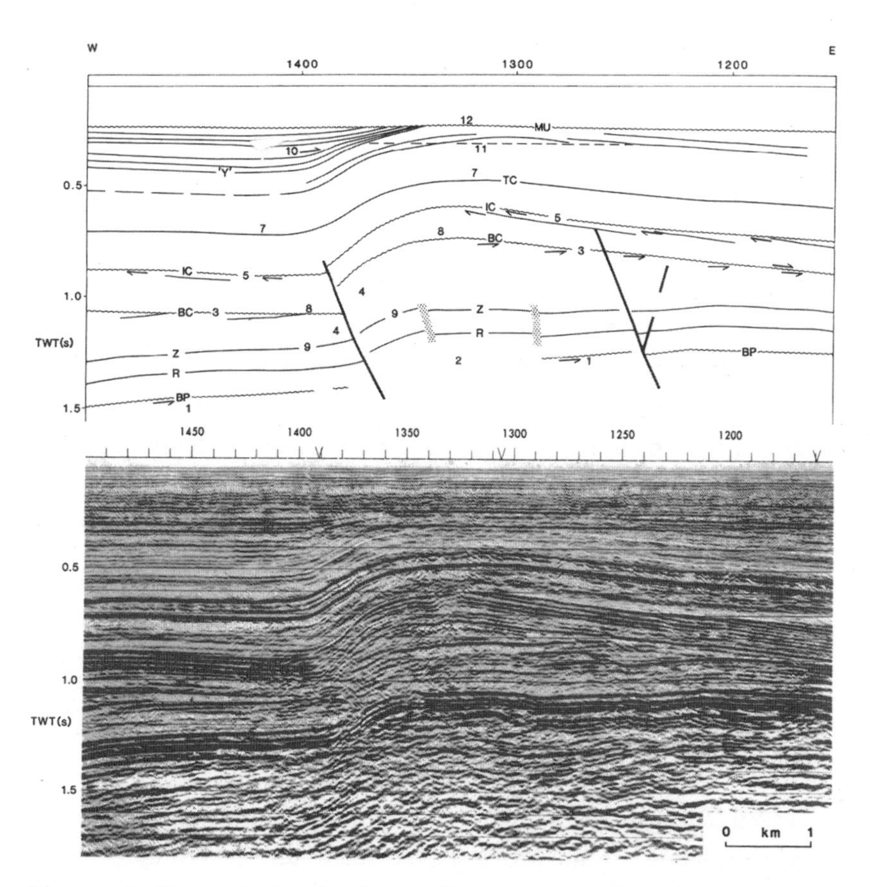

Figure 36. Deep seismic reflection profile cross section through a hidden fold

regular multilayered folds, the product of hundreds of repeated underlying fault displacements, without the accompanying river or wave erosion that would 'disfigure' the same fold above sea level.

The 50-kilometre-wide North Celtic Sea Basin and the faults within it run parallel to the WSW-trending southern coast of Ireland, at its closest, only 50 kilometres offshore. Although the basin is now completely underwater it shares a tectonic history with that of southern England.[21] Triassic and Jurassic age rifting was followed by slow subsidence and more than a kilometre of chalk deposition. In the Eocene the basin was squeezed, forming broad shallow-dipping anticlines, similar to the Weald Basin. Fifty-three separate anticlinal folds above blind reverse faults have been identified, three of which proved to be traps for commercial quantities of natural gas.[22] The total uplift on some structures reached 1,000 metres. Formerly a series of downs and vales like Sussex, the hills were eroded by the Atlantic waves and then sank down, so the margins were overlain by undeformed 40 Ma sediments. This identifies the compressional tectonics as 'Pyrenean'.

Somewhere between St David's Head, Pembrokeshire, and Carnsore Point, Ireland, lies the north-west end of the Sticklepath Fault.[23] To the west run the anticlines of the North Celtic Sea Basin, revealing NNW-directed Eocene squeezing.[24] The oldest strata from the base of the Petrockstow Basin contain the isotope record of the climate warming episode dated to 52.5 Ma, within 5 million years of the last eruption from the Lundy Volcano and contemporary with the resumption of the Africa–Eurasia collision. The Sticklepath Fault was moving through the Eocene, while the blind reverse faults beneath the North Celtic Sea Basin were busy sculpting anticlines.

In the middle of the oceans, a strike-slip 'transform fault' connects two offset lengths of spreading ridge. In a similar way, the Sticklepath Fault behaved as a 'collision transform', connecting the crustal shortening of reverse faults and monoclines from the Weald, Wessex, and Channel faults through to the crustal shortening of reverse faults and monoclines beneath the North Celtic Sea Basin.

Further to the west the band of Eocene reverse faults can be chased along the southern end of the Rockall Trough, to the west of Ireland, eventually connecting into the North Atlantic spreading ridge.

Spain moved with the African plate up to 37 Ma,[25] and thereafter switched allegiance to be part of the Eurasian plate. The Pyrenees were no longer the plate boundary front line. These noble forested mountains, perhaps only half the height today of their culmination at the end of the Eocene, still manifest their deep tectonic roots in a profusion of thermal springs.

★

The Sticklepath Fault, it turns out, had more than one role in the Eocene. To the east of the north-west end of the Sticklepath Fault a deep Eocene rifted trough (the St George's Channel Basin) opened.[26] Even while the Pyrenean collision was dominating the tectonics, around the jogs on strike-slip faults it was still possible to find localized extension (Fig. 37). The St George's Channel Basin was a yawning 'pull-apart', opened where the dextral strike-slip displacement on the Sticklepath Fault jumped to the north-east to be picked up by dextral displacement on the NNW-trending Codling Fault.

The intervening basin, which accumulated a kilometre thickness of river delta sands, layers of lignite, and occasional marine clays, was controlled by extension on the NE-trending Bala Fault, which runs nearshore parallel to the Cardigan Bay coast. The basin was sinking and sediments accumulating through the Eocene and into the Oligocene (50–30 Ma).[27] As with the pull-apart basins on the Sticklepath Fault, the ages of these sediments reveal when the Codling Fault was also moving.

Like a tear through the perforations in a sheet of stamps, the paths of both the Sticklepath and Codling faults were determined by the legacy of the hot, weak crust beneath the volcanoes. Where the Sticklepath Fault heads offshore to the north-west of Devon it is making a beeline for the edge of the Lundy volcanic centre and is

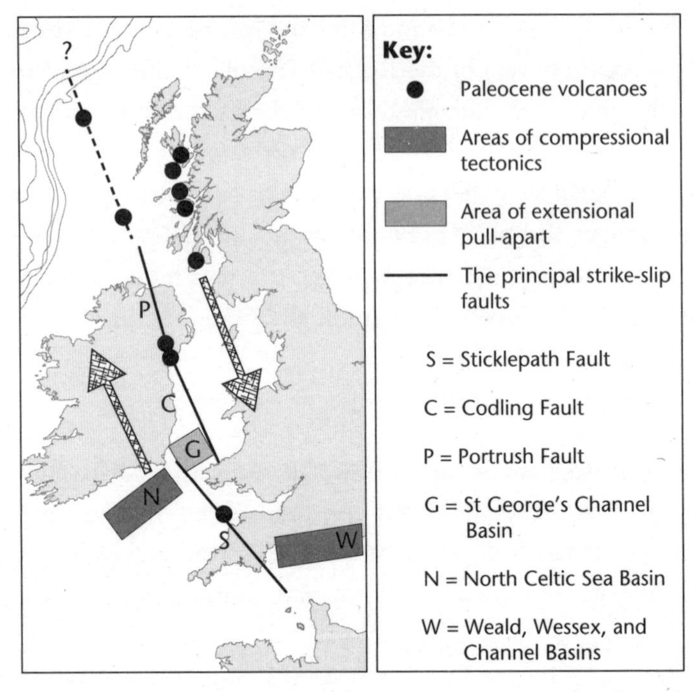

Figure 37. Eocene 45 Ma tectonics through western and southern Britain

joined by a parallel fault on the other side of the volcano, 15 kilometres to the south-west.

The NNW-trending Codling Fault runs underwater 30 kilometres east of Dublin, cutting the Kish Bank Basin, where it surgically offsets the Lambay and Dalkey faults by 4 kilometres of dextral displacement.[28] Continuing north, the Codling Fault aims for the heart of the Carlingford and Slieve Gullion volcanoes,[29] where the faults are braided into four fractures, together achieving at least 4 kilometres of dextral offset (Fig. 38). The stumps of these volcanoes provide the best place to experience traces of this great braided strike-slip fault zone. Intrusions are visibly dextrally offset. You can walk the line of a fault and find a rocky knoll on one side repeated hundreds of metres, and in one case 2 kilometres, farther along on the other side. To the north beyond the volcanoes the fractures converge into two NNW-trending faults, 5 kilometres apart.

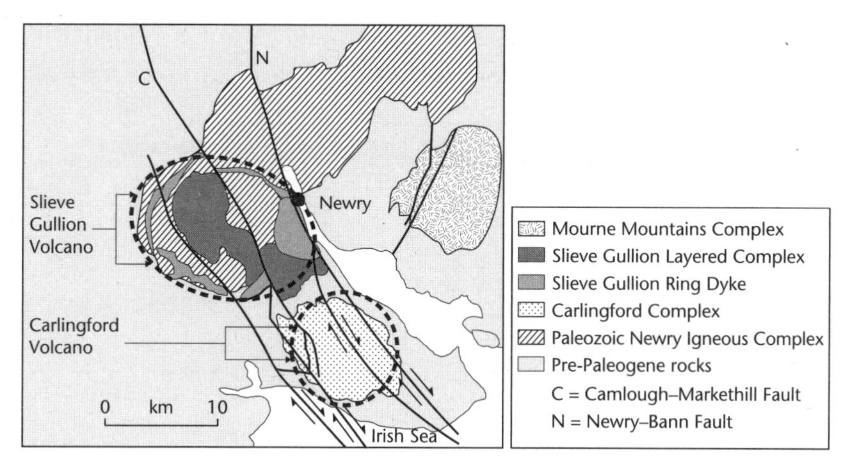

Figure 38. Dextral strike-slip NW–SE faults cutting through the Carlingford and Slieve Gullion volcanoes

There are no Eocene age pull-apart basins in north-east Ireland and so there could not have been any significant right-hand fault jogs on the NNW-trending faults. Around the Carlingford Volcano we surmise Eocene movement concentrated on the easternmost NNW-trending Newry Fault (which exhibits a dextral offset of 2.5 kilometres on the 400 Ma Newry Igneous Complex), before linking via the Bann Fault to the 60-kilometre-long Portrush Fault.[30] On the coast to the west of Portrush town there is a famous sill, to the east of the town lavas and dykes. This highlights the Portrush Fault was a 'stress frontier' 59 million years ago. The Portrush Fault follows the western 'volcanic ley line', we can presume, heading offshore for Blackstones Volcano and probably on through the St Kilda Volcano.

The composite Codling–Newry–Bann–Portrush Fault is at least 300 kilometres long, and with the projected extension to the north could be at least 500 kilometres long. Even then, several kilometres of horizontal displacement cannot simply terminate, but must ultimately connect into the Norwegian Sea plate boundary.

Having seen the NNW alignment of the offshore Codling Fault helps explain a mystery raised at the start of this story: why through its relatively short length did the Sticklepath Fault generate two prominent pull-apart basins? The geology of Devon is ridden with NW–SE trending faults formed at the culmination of Variscan 300-million-year-old tectonics. Yet as revealed by Codling, the orientation of this composite mega-fault wanted to be NNW. This required the Sticklepath displacement in Devon to jump to the right from one NW–SE fault to its neighbour. Each jump gave us a pull-apart basin, while the overall displacement trended more northerly.

7

The great lignite hunt

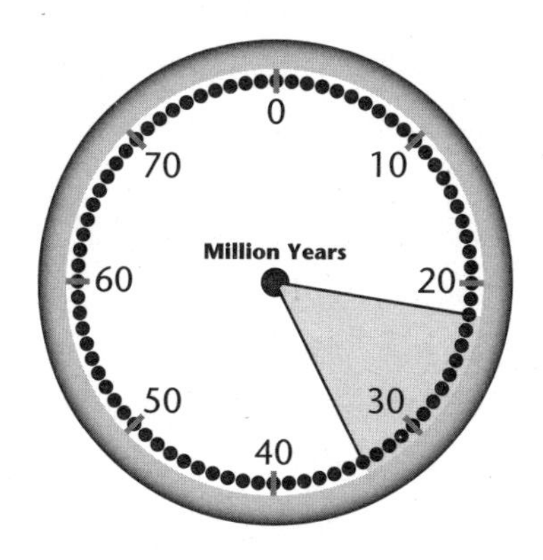

In the middle of the Eocene, around 45 Ma, even while the African–Eurasian plate collision at the Pyrenees was raising mountains, farther east a great NNE-trending crack started widening near what is now the border between Germany and France, from Basel to Baden Baden. This initial rift became a lake which, as the floor sank, filled with 900 metres of freshwater marls. By the Oligocene (around 33 Ma) the rift had expanded farther to the north and was now a 400-kilometre-long widening crack running from Basel to Mainz[1] where 1,800 metres of mud and sand filled the sinking lake floor.

The German name for a rift valley is 'graben', meaning a ditch. Today a great river, the Rhine, sourced in the snowfields and glaciers of the Alps, runs north along this rift valley, and the rift is known as the Rhine Graben, but there was no such river for most of its history. Over the initial 20 million years, the 50-kilometre-wide rift valley extended and widened by more than 5 kilometres.[2] For most of that time the rift was filled with lakes, like the rift valleys of East Africa. By 25 million years ago, the rift had become an embayment connecting the North Sea with the Mediterranean, before the graben got blockaded by the advancing Alps.

What was happening takes us back to the 'crack in the wall' lessons from the start of this story. No rift can simply come to an end. The opening has to shift to another crack, another rift.

At the southern end of the Rhine Graben the rift jumps 120 kilometres to the west and the opening continues south on the Bresse Graben, a rift valley that extended by 5 kilometres from 45 to 30 Ma (see Fig. 39). The town of Besançon is founded on the complex mesh of faults which links the two graben. On the west side of the graben there is Dijon in the north, Chalon-sur-Saône in the centre, and Lyon in the south: this is 'foodie' terrain. Further south the E–W stretching of the Bresse Graben passes into a series of parallel rifts continuing to the Mediterranean coast. Much of the eastern side of the Bresse Graben was subsequently overridden by the advancing Alpine mountain front.

Around 30 million years ago the widening of the Bresse Graben stopped, and crustal stretching shifted another 150 kilometres farther to the west onto the N–S trending rifts of the Limagne Graben, which filled with 2,000 metres of sediment. (Limagne is derived from the Latin 'Lacus Magnus'. In Roman times the fertile plain would flood after heavy rain.) Vineyards stretch along the western slopes, where the principal fault that controlled the subsidence was situated. Overall the basin has stretched east–west by 2–3 kilometres, continuing to open into the late Oligocene, 25 Ma.

The northern end of the Rhine Graben rift is another great crack that cannot simply come to an end. Most of the rift-opening passes to the north-west into the Netherlands along WNW-trending faults

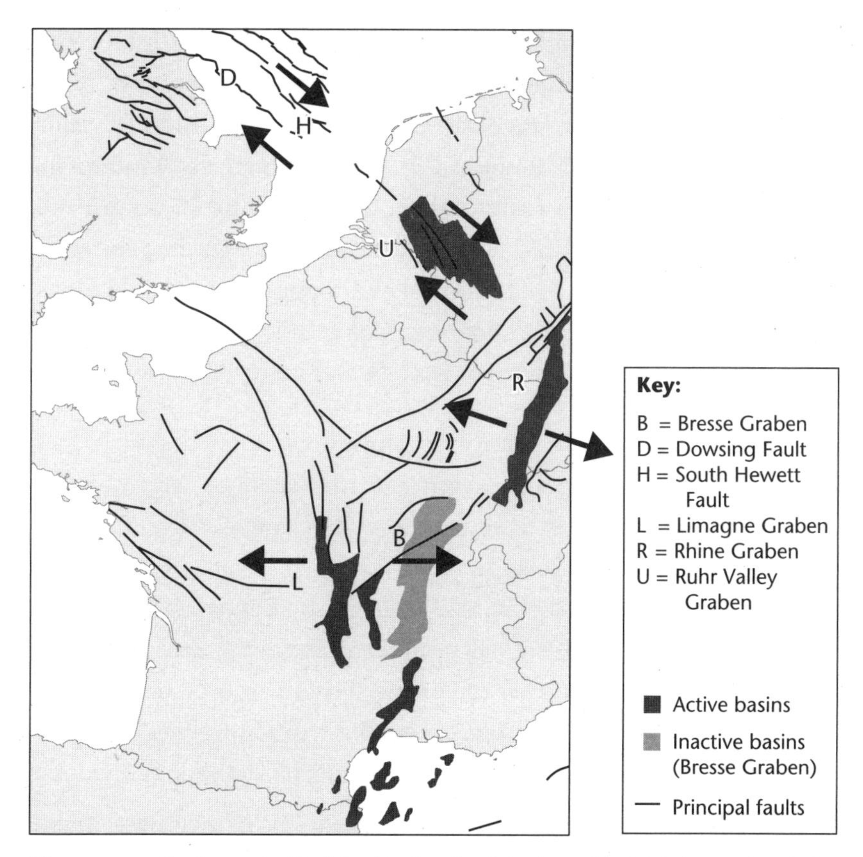

Figure 39. Faults and rifted basins involved in 'transferring' the ends of the Rhine Graben rift opening

with horizontal displacement. Where the faults trend more to the NNW, the sideways movement accompanies 'pull-apart' extension, creating deep Oligocene basins, which filled with sediment. The 'dextral' sideways movement continues into the south-west North Sea along the WNW-trending South Hewett Fault, which comes to within 100 kilometres of the north-east corner of Norfolk. Movement passes into the NW-trending Dowsing Fault where the trail is obscured by the mass mobilization of thick layers of salt, from the early history of the North Sea, when it was a landlocked basin in the middle of a great continental desert.[3] At a few locations faulting can be seen passing onshore, as along E–W fractures through the tortured chalk cliffs at Selwicks Bay close to Flamborough Head.[4]

We think this 'tectonic highway' continues into northern England, through the Cleveland Basin north of the Yorkshire Wolds. The WNW-trending Craven faults and the NW-trending Pennine Fault imitate the faults in Holland and offshore and may continue the sideways movements. Both faults control the topography, from Malham Cove to the 'Alston Block', and look geologically 'young'.

★

It was the results from gravity and magnetic surveying that first suggested something strange about the mountainous west coast of north Wales. Immediately offshore, the pull of gravity falls off steeply, implying thick layers of low-density rocks.

Founded in 1283, Harlech Castle was built by Edward I on a coastal bedrock prominence. Originally supplied from the sea through a water-gate, over the centuries the primary enemy proved to be blown sand rather than the Welsh. Scoured by the wind from the river estuary at low tide, dunes arrived to choke Harlech's harbour and extend a sandy peninsula.

By 1966 discoveries of huge gas fields, first in the Netherlands and then the southern North Sea, led the UK government to take the initiative in mapping the geology beneath the seas around Britain. Campaigning by the Head of Geology at the nearby University of Aberystwyth succeeded in prioritizing the exploration of Cardigan Bay. In November 1967 drilling started at the Morfa Dyffryn sand dunes to the south-west of Harlech. At great expense the borehole was to be carefully cored.

As they started drilling, the geologists had no idea what the borehole would reveal. Inland of the coast the youngest rocks were grits and slates, almost 500 million years old.

First the borehole sampled recent blown sand, before coring tens of metres of glacial 'boulder clay'. But next the geologists pulled up cores of Cenozoic age sand, clays, and lignite seams, which continued for the next 600 metres. To the core-loggers, this was a total surprise.

There was nothing like this anywhere in Wales. Perhaps the nearest equivalent, one geologist suggested, was in the rifted basins along the Sticklepath Fault.

Beneath the clays and lignite, there was a big gap in the geological sequence, completely skipping the Cretaceous and much of the Jurassic. The whole region must have been uplifted and eroded, losing a great thickness of chalk, in the general Paleocene uplift of Britain. Down the borehole, stratigraphy was resumed with mudstones, sandstones, and limestones, containing characteristic Early Jurassic 'Lias' age (180–200 Ma) ammonite fossils.

Between this huge pile of Cenozoic and Jurassic sediments, and the tough Cambrian sandstone grit mountains inland of Harlech Castle, there had to be a fault following the former north–south coastline. At Harlech there was at least 4.5 kilometres of vertical offset on this 'Mochras Fault', perhaps the largest of any fault on-land in Britain. (The newly identified fault soon became notorious, claimed as a source of mysterious lights and flying saucers.)

The Cenozoic sediments encountered in the borehole had been deposited by meandering rivers with swampy margins, snaking across a broad flood plain and flowing into a large lake to the south. Fossil pollen placed the age of these sediments to the mid-Oligocene to Early Miocene (30–20 Ma).

We know about the swamps because of the thick seams of lignite encountered in the borehole. Lignite is compressed peat, squeezed by the weight of all the overlying sediments. Peat forms from the remains of dead trees and plants sunk in swamps and lake deltas, away from the open sea and dissolved oxygen.

The lignite reminds us that for 60 million years, since Britain and Ireland first emerged out of the sea, right until the arrival of the Arctic cold that preceded the ice sheets, the land was thickly forested. Close to the west Atlantic coast, tangled, moss- and fern-rich rainforests cloaked the hills, whereas in the rain-shadowed east, woodlands were drier and more open. And yet there is precious little trace of these forests. A tree would sprout, grow up to the canopy, be home to mammals and birds, and eventually die, tumble to the ground, and completely rot away, leaving nothing behind. Very rarely, as in a vertical trunk withstanding a lava flow, do we have evidence of the mighty trees. The timber itself only survived when the tree fell into the stagnant water of a swamp, whereas wind-blown tree pollen might be preserved in a former lake-bed: the smallest vestige of the once-magnificent forests that stretched across Britain, horizon to horizon.

Long before the Ice Ages it took tectonics to create a lake. Thirty million years ago Britain and Ireland were still an elevated landmass. Any isolated area of tectonic subsidence would become an inland lake surrounded by swampy forested shorelines. In such a situation the key hallmark of a past lake basin is lignite.

★

By the end of the Oligocene the tectonically sculpted lake basins had multiplied (Fig. 40). Travel back in time 25 million years and, seen from the sky, the landscape of the British Isles landmass is more beautiful than at any time over the past 60 million years. A necklace of emerald jewels spans from Brittany to the northern Hebrides. Each lake is fringed with thick forest. In each the original floor has sunk hundreds of metres while the basin has filled with sands, clays, and lignite. These are the Celtic lake basins. Some were fed by broad rivers and intricate deltas.[5] Some were lakes without inlets. Some were swamps with meandering channels. All of them have laid down thick seams of peat that has been stewed and squeezed to make lignite.

We will start in the south. Through Brittany[6] the NW-trending Quessoy–Vallet fault zone[7] is marked by a chain of small Oligocene 'pull-apart' étangs (small lakes).

Along the path of the Sticklepath Fault through Devon we have the rhombic outline of the Bovey lake and the cigar-shaped Petrockstow lake.

North from Devon, the next lake basins are today underwater. The Stanley Bank lake lay to the east of the ruined stump of the Lundy Volcano, dammed by the Sticklepath Fault. Shallow cores reveal layers of the talisman lignite, up to 5 metres thick and of middle Oligocene age.[8]

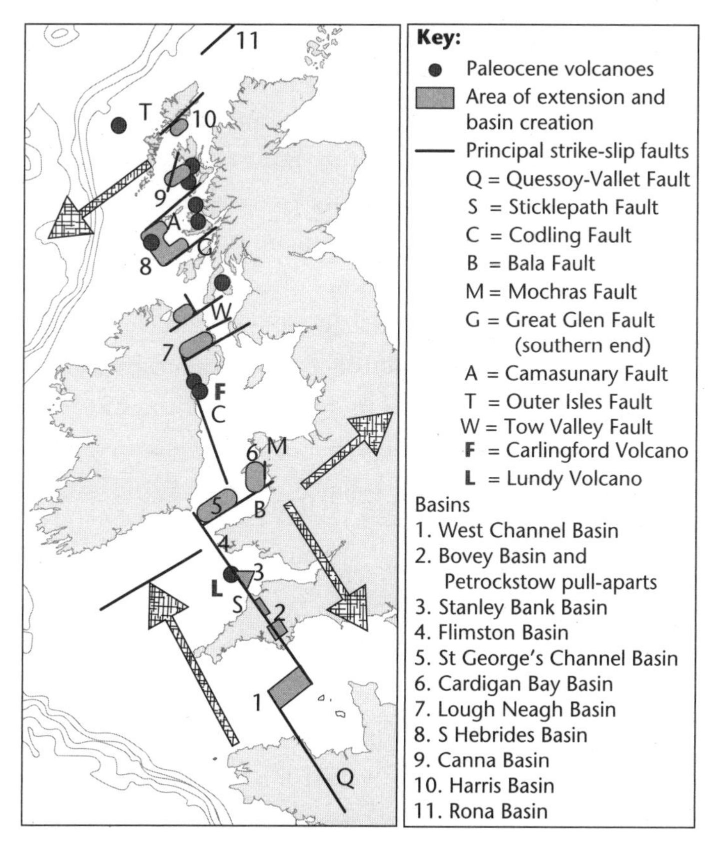

Figure 40. End Oligocene (25 Ma) tectonics of western Britain

A branch off the main Sticklepath Fault intersects the stark cliffs at the south-west tip of Pembrokeshire at Flimston. Fissures and clefts opened along a 100-metre-wide, NNW-oriented pull-apart, the site of late Oligocene pools[9] that became filled with lignite, sands, and clays, formerly quarried for making pottery.[10]

The biggest swampy lake of all was in the heart of St George's Channel, a pull-apart first formed in the Eocene, covering more than 4,000 square kilometres.

For the Cardigan Bay Basin, we have already identified late Oligocene and early Miocene swamps and rivers from the Mochras borehole.

Heading north, the next lake is back 'on land'. The Codling Fault, sister to the Sticklepath Fault, was active at this period. We can surmise displacement took a more western track through the Carlingford Volcano than in the Eocene, continuing along the Camlough Fault to bring 2 kilometres of dextral displacement of the Slieve Gullion ring-dyke granites before aiming for the south-west corner of the Lough Neagh basin. Now in the more extensional tectonics, a jump in the line of faults has created a pull-apart and plentiful lignite.

Lignite was first found on the shores of Lough Neagh in Northern Ireland in 1757,[11] at Crumlin to the east and Coagh to the west.[12]

Today's shallow lake, a product of glacial erosion, has flooded the tectonic lake which existed here in the middle of the Cenozoic. The waters of Lough Neagh were once believed to cause petrification—turning tree branches into stone—but these are fossils dating back to the late Oligocene (25 Ma).

The original tectonic lake covered 550 square kilometres and is filled with up to 500 metres of sediments overlying the lavas. Many cores have been drilled exploring for lignite. Near Crumlin, lignite seams are 22 metres thick, with some sequences revealing almost pure lignite for 250 metres. The tectonic lake evidently had no significant sediment-laden streams flowing into it. The lake was low-lying and frost-free, surrounded by luxuriant forests of ferns, conifers, palms, and swamp cypress. Fossil salamander teeth fall out of mudstones near Coagh[13] along with gastropod shells.[14] It seems Oligocene fish never made it to this isolated water body. Through the 1980s there were plans for a power station to burn 200 million tons of lignite, excavated from an area to the north and east of Crumlin. By damming the edge of the lake another 350 million tons of lignite would be accessed. Fortunately, action on emissions put paid to these polluting schemes.

Boreholes drilled at Ballymoney, on the north side of the NE–SW Tow Valley Fault, 40 kilometres north of Lough Neagh, also found plentiful late Oligocene lignite, in seams from 20 metres to 200 metres thick,[15] accumulated in swampy compartments bounded by NNW-striking faults.

Beyond the northern tip of Antrim, our lignite lake chase has to return underwater, employing seismic reflection profiles, sea-floor sampling, and coring.[16]

The next deep lake basin, assumed to be filled with sands and lignite, has only been identified from seismic profiles, in the southern Sea of the Hebrides. The basin is at least 30 kilometres across and 800 metres deep, involving 500 metres downward displacement on the NE–SW Camasunary–Skerryvore Fault, and another 500 metres on the offshore extension of the NE–SW Great Glen Fault. The downwarped basin extends over part of the stump of the Blackstones Volcano, which is probably why this great volcano is entirely underwater.

Shallow cores into the Canna Basin to the west of Skye revealed lignite, dated as late Oligocene. The former lake basin contains up to 800 metres of these 'continental' sediments, overlying a thick succession of basalt lavas.[17] The tectonic downwarping that formed the Canna Basin continues into western Skye. Offshore the base of the lavas has dropped to 1,500 metres below sea level from displacement on a big NNE-trending extensional fault on the west side of the basin.

Heading north, a lake basin was seen in seismic reflection profiles up against the Outer Isles Fault, on the eastern edge of the island of Harris, with the sampled lignite-rich sediments dated again as late Oligocene.[18] The final lignite haul was found by chance in a borehole (77/7) drilled 80 kilometres north of Cape Wrath, close to the storm-swept island of North Rona,[19] highlighting that the land formerly spanned far to the north and included all of Orkney and Shetland.

Having admired this garland of Celtic lake 'jewels', we can ask: what caused all these lake basins to form? Plate tectonics teaches us to expect consistent tectonic geometry over large distances.

Between the north-west continuation of the Sticklepath Fault and the south-east extension of the Codling Fault, there is a jog of about 120 kilometres, far wider than a classic pull-apart. In the Eocene, the intervening St George's Channel Basin behaved like a giant pull-apart, as we noted earlier, with the basin opening between the bounding faults. Yet through the Oligocene the NE-trending Bala Fault, which ran along the south-east side of the basin, transitioned from being an Eocene extensional normal fault into moving as a sinistral strike-slip fault. We know this because a prominent left-handed jog in the Bala Fault then became the site of the N–S Mochras Fault, which, at the end of the Oligocene, itself underwent 'pull-apart' extension to create the local 'Cardigan Bay basin', first encountered in that 1968 borehole.

By the end of the Oligocene the tectonics of the Celtic Lake basins involved two sets of strike-slip faults, one dextral oriented NNW, the other sinistral running NE, almost at right angles to one another. In combination they opened a series of extensional basins, each much bigger than the classic cigar-shaped Petrockstow pull-apart basin (Fig. 40).

The two sets of strike-slip faults helped accommodate the stretching and transfer it from basin to basin. Through the Hebrides the Celtic basins were controlled by some of the same transform faults that interlinked the lava fields. The NE Camasunary Fault on the island of Raasay has achieved 750 metres of vertical displacement since the time of the great volcanoes, but also likely carried horizontal strike-slip displacement.[20]

The principal crossroads for the two orientations of fault systems was in the centre of Northern Ireland.[21] (The two sets of faults can't actually cross one another and both be active at the same time, but one set can be juxtaposed against the other.) The Lough Neagh basin is, from one perspective, a pull-apart on a jog between NNW dextral faults. However, the faults mapped within the basin all run NE[22] and the basin might better be considered as forming at a left-handed jog between NE–SW sinistral faults.

To the north of Lough Neagh, the NE–SW Tow Valley Fault is the predominant fault, whereas NNW–SSE faults, including the Portrush Fault, active in the Eocene, terminate up against it. Likewise, the southern Sea of the Hebrides basins are also controlled by N–S faults, we assume to be principally strike-slip, and the Canna Basin by a big NNE-trending normal fault.

★

To understand what was happening 25 million years ago we need to expand horizons and find the big-picture, plate tectonic perspective.

Through the first 10 million years of the Norwegian Sea spreading ridge (from 55 to 45 Ma), separating Greenland from northern Europe, the continents moved apart north-west to south-east. At the same time the Labrador Sea to the south-west of Greenland (separating Greenland from North America) was opening north-east to south-west. As the Labrador Sea spreading dwindled and died, the spreading in the Norwegian Sea shifted to carry all the movement between North America and Eurasia. To do this, instead of opening north-west to south-east, the spreading ridge transitioned to opening east–west.

This caused a problem with the transform faults offsetting sections of the Norwegian Sea spreading ridge. These transform faults guided horizontal movement. To perform their function, like great rails, they had to be oriented precisely in the direction of opening. What would happen to the transform faults if the opening direction should change (Fig. 41)?

Look along the axis of the Eocene age NE-trending Norwegian Sea spreading ridge. The ridge was offset by intervening transform faults, both to the left and to the right. To the west of the modern Faroes, at a transform fault, the spreading ridge stepped to the left. Farther south and to the west of the Shetlands there was a prominent transform fault where the spreading ridge stepped to the right, keeping to the north and west of the Rockall Bank.

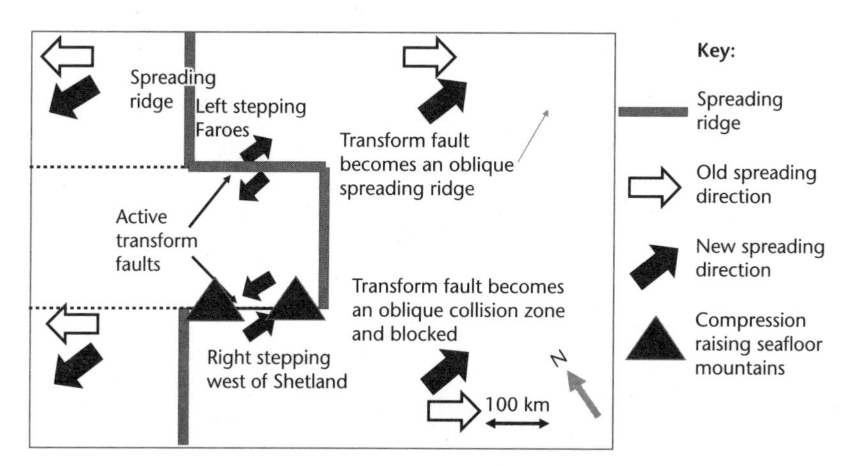

Figure 41. How the transform fault west of Shetland became a collision zone after a rotation in spreading directions in the Norwegian Sea

For the 'Faroes' transform fault with its left-hand offset, the shift in the spreading direction meant the NW-oriented transform fault became a zone of E–W extension and oblique spreading, with the underlying mantle rising to fill the gap and create new ocean crust.

For the 'west of Shetland' transform fault with its right-hand offset, the two sides of the fault were now jammed in oblique collision, grinding into one another.[23] This oblique collision created enormous stresses and over several million years forced up two tightly folded ridges west of the Shetlands, the 200-kilometre-long and 2-kilometre-high Wyville Thomson Ridge, which absorbed 4 kilometres of crustal shortening, and the parallel but shorter Ymir Ridge. Both ridges trend close to NW, along the line of the former transform fault, situated at the northern end of the Rockall Trough. Ocean crust and the underlying mantle are strong and this locked transform fault put a brake on sea-floor spreading in the Norwegian Sea[24] (see Fig. 42), which reduced down from 20 millimetres per year when spreading had started in the Eocene, to less than 5 millimetres per year by 30 Ma, in the Oligocene.

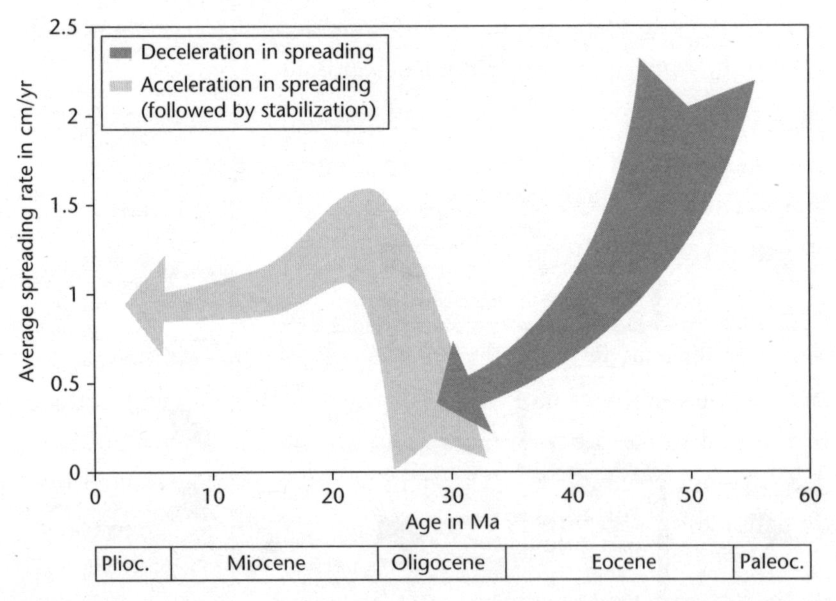

Figure 42. Changes of spreading rate in the Norwegian Sea (caused by the blocked transform fault brake on spreading)

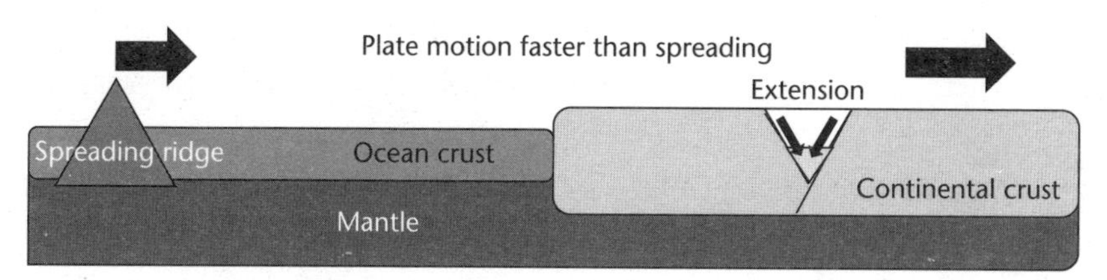

Figure 43. When the plate is moving faster than the opening at the spreading ridge we get regional extension

Meanwhile, plate tectonic forces continued separating the Eurasian plate from the North American plate. Yet, if all the separation was not happening at the spreading ridge, some of it would have to happen within the plate (Fig. 43).

Imagine tugging at a thick sheet of wet paper. Where there was a pre-existing linear score mark on the paper, that is where the paper starts to split—as along the Rhine Graben (Fig. 44). Where there were some pre-existing holes in the paper, that is where gashes start to open—this was the situation in south-east France. Through western Britain the stretching had become concentrated around the hot, weak crust beneath the Paleocene volcanoes, and was accommodated by two orthogonal sets of strike-slip faults, one among them being the Sticklepath Fault. Take a loose set of square floor tiles, laid without gaps on the floor. Offset the tiles a short distance along one axis, and they are still tightly packed. Offset a short distance on both axes at 90 degrees and you create a pattern of openings.

⋆

For 30 million years, through the Eocene and the Oligocene, the faults to the south, east, and west of Britain were buzzing with activity. In the mid to late Eocene, from 45 to 40 Ma, there were pulses of deformation arriving from the Pyrenean collision zone, leaking into the latitudes of southern England, and passing all the way west into the Atlantic. At the same time most of Ireland (with Cornwall) was slipping north along the NW and NNW strike-slip faults that ran from Brittany to beyond the Hebrides.

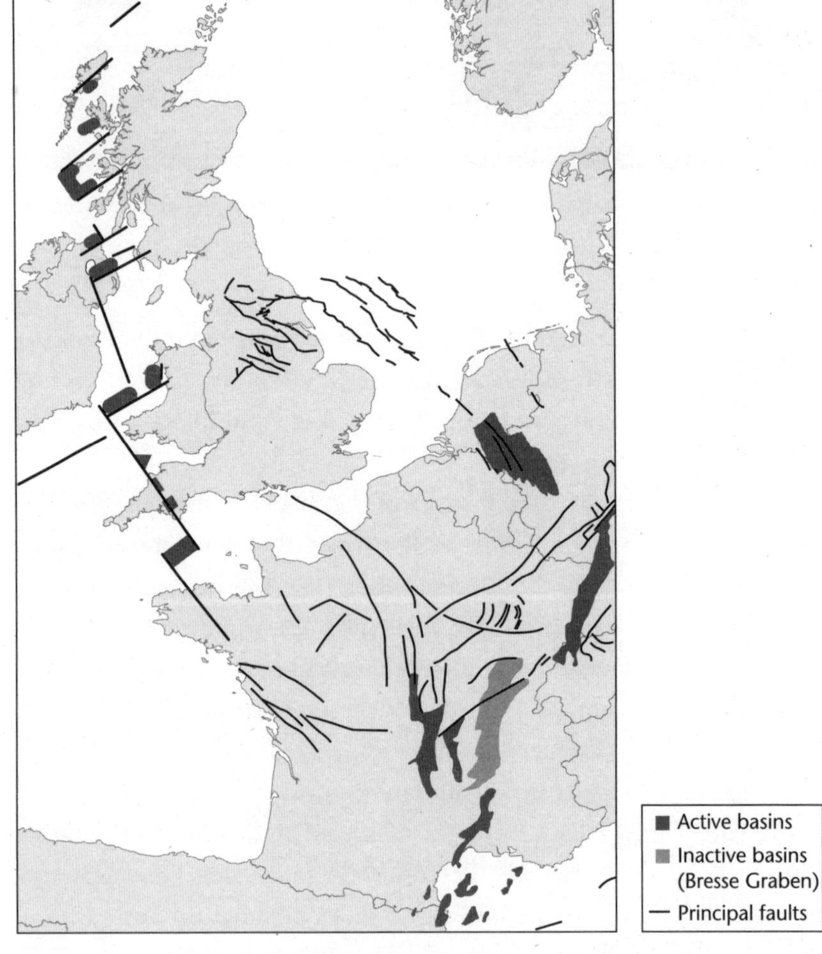

Figure 44. End Oligocene tectonics across western Europe

Next came early Oligocene (30 Ma) horizontal displacement, accommodating the opening of the Rhine Graben, on NW and WNW faults running from the Netherlands past East Anglia into northern England.

Towards the end of the Oligocene (25 Ma) the pattern of faulting connected a series of rifted Celtic basins, from Devon through Cardigan Bay to the Hebrides. Our 'crack in the wall' lesson highlights that all these interconnected Celtic basins must have been opening at the same time.

The crossroads of the two principal fault directions was in northeast Ireland, which had been a principal centre of Paleocene volcanics and dyke intrusion.

In each of these tectonic episodes, fault movements were sporadic and catastrophic. As with active faults on plate boundaries today, a sudden fault rupture lasts tens of seconds, generating the strong vibrations we call earthquakes. Then for hundreds, or thousands, of years the fault resumes its dormancy, while the stresses slowly rise that will eventually cause it to break again.

What was the earthquake activity across the Weald Basin 40 million years ago? Did each blind thrust fault break individually so there was a Dorking earthquake and then a Reigate earthquake? It seems more likely that, as around Los Angeles today, the ruptures jumped from one fault segment to the adjacent one. The faults were shallow and the vibrations would have been locally strong and quite damaging, like the 1994 Northridge, Los Angeles earthquake, from the rupture of an underlying blind thrust fault. If these earthquakes resumed today, there would soon be no chimneys left in south Surrey and north Hampshire. The timber-frame buildings might prove more resistant to strong shaking.

Living in southern and western Britain, from 45 Ma down to 25 Ma, would have been like living in modern Turkey, with the potential for infrequent devastating earthquakes. And at the end of this period, the earthquake activity was by no means over.

8

The Miocene dark ages

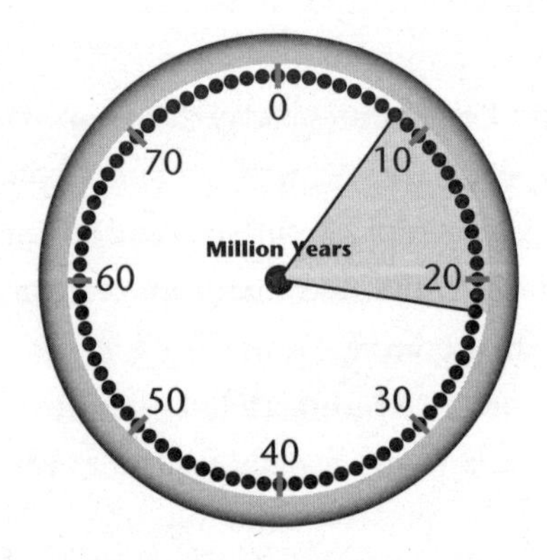

For hundreds of years, villagers across Britain dug into the nearby bedrock for building-stone, out of which they erected structures in which to live, keep their animals, sharpen their tools, drink their beer, and worship their God/gods. To do this they excavated rock formed within all the geological ages.

Building-stone is strong, insulating, cleaved, or bedded but, above all, heavy. When materials had to be carried by wheelbarrow or cart, wherever possible stone would be locally sourced. (Only a rich bishop could afford to have Caen limestone transported by barges to build his cathedral.) A pit soon grew into a quarry for the use of the whole village. Anyone planning to build a house could go and chisel their own rubble, or pay a stonemason for the digging.

Across Britain you can map the local geology from older, unplastered exterior house, barn, and church walls. The stone in these walls is typically far better exposed than the quarry from which it was sourced. Every geological age, from Precambrian to Oligocene, can be found in the outer walls of some British building. (I have seen house walls in Sussex and Dorset reflecting the underlying diversity, interleaving three geological ages in the stonework.)

Until, that is, we arrive at the Miocene. Find me a British Miocene stone house or church!

You might think sediment layers would become more abundant the nearer they are to the present day, with less time for erosion to have worn them away?

It is not simply that the whole record runs out. Before the Miocene, Oligocene building-stone includes the shell-packed 'Quarr Stone' from the north coast of the Isle of Wight, used so prodigiously in medieval abbeys, castles, and cathedrals that there is none left at its source.

The Miocene is followed by the Pliocene and we are back with building-stone. Both the fourteenth-century Wantisden and Chillesford churches in Suffolk have towers built out of sawn blocks of Pliocene 'Coralline Crag' (rich in Bryozoan—microscopic animals that form colonies in the shape of mineralized fans, bushes, and sheets). We even have Pleistocene buildings, as at Ufford Church where rough blocks of ironstone have been arranged into herringbone patterns.

It is not that the Miocene epoch was short. It spanned 60 per cent longer than the preceding Oligocene and more than six times longer than the subsequent Pliocene. And yet the Miocene has left (almost) no trace across Britain. The Miocene is Britain's geological dark ages.

In most geological periods, land has risen and land has sunk. Typically it is only in the areas that subsided that evidence is preserved, where sediments or lava flows accumulated. In geology, when nothing is left behind, that tells us that no land sank during that time.

Yet even without changes in land level, sea levels in the Miocene were higher than today. The Miocene ice caps in Antarctica and Greenland were a lot smaller. Higher sea levels mean more land is inundated, leaving beach deposits that should be exposed when the levels fall again. What happened to these coastal sediments across Britain?

The on-land legacy of Miocene age material is truly meagre. The only identifiable British Miocene sediments were laid down out of reach of the sea, blown by the wind into hollows, or carried by floods into caves.

In northern Anglesey, in crevices in the limestone, black clays are rich in pollen grains and spores identified as Miocene forest and bush species, reflecting a warm-temperate climate.[1]

In the southern Peak District around Brassington,[2] sands and clays accumulated in karstic depressions in the Carboniferous age limestones.[3] Plant spores in the clays are of Upper Miocene age. The surrounding woodlands were a mixture of Asian species like cedar, hemlock, sweetgum, the Japanese umbrella pine, sweetleaf, and Japanese cedar alongside familiar British species like alder, spruce, and hazel. Sea thrift and sea lavender suggest that the sea was not far away, although today this site is 330 metres above sea level. The vegetation indicates a mellow mean annual temperature close to 16°C, as compared with 10°C today.

At the village of St Agnes on the north coast of west Cornwall,[4] a wind-blown deposit containing Miocene age fossil spores was found next to an abandoned cliff-line.

★

The name coprolite (meaning 'turd stone') was coined by the irreverent Reverend William Buckland, Dean of Westminster, the most formidable geologist of the early nineteenth century. On a geological excursion to Lyme Regis to see the fossils exposed along the cliffs, he found knobbly stones that he believed to be petrified dinosaur dung.

Even while the name stuck, it turned out these turd-shaped nodules had never been excreted, but had simply recrystallized from fossil teeth and bones on the sea floor. Coprolites were rich in the mineral apatite—a source of phosphate and a valuable fertilizer—but the apatite was not readily broken down in the soil. However, once dissolved in sulphuric acid the apatite assumes a soluble form ('super-phosphate') easily absorbed by plants.

There was big money to be made in improving agricultural yields. John Bennet Lawes, a Hertfordshire landowner who had coined the name 'super-phosphate', tested alternative manures on his crops. In 1839 he set up the Lawes Chemical Manure Company, and in 1842 obtained a patent for super-phosphate. He identified a plot at Deptford Creek in the Thames Estuary and built a factory to manufacture 200 tons of super-phosphate each week.

Meanwhile, after declining to train to become a doctor in Edinburgh, in 1828 nineteen-year-old Charles Darwin came to Cambridge ostensibly to study theology. He befriended the Reverend John Henslow, former mapper of the geology of Anglesey and now Professor of Botany and Mineralogy at St John's College. Henslow encouraged Darwin to become a naturalist.

Henslow's brother-in-law had been invited to take up the post of official naturalist on a two-year, round-the-world naval survey. After declining the invitation he passed the offer on to Henslow, whose wife dissuaded him from applying. Henslow then suggested to the bachelor Darwin that he apply for the post of geologist/naturalist on the survey vessel, HMS *Beagle*. This is how opportunity arrives.

In 1837 Henslow was assigned by his college a well-paid 'living' of £1,000 a year in the parish of Hitcham, Suffolk. Most college fellows in this situation simply turned up a few times a year to deliver a sermon while appointing a deputy. Two years later, Henslow came to stay at the nearby coastal village of Felixstowe and decided to move into his parish, with the ambition to improve the lives of his parishioners.

In a cliff at Felixstowe, Henslow found blue-grey (Eocene) London clay overlain by red shell-rich Crag deposits (of early Pleistocene age). At the 'unconformity', or 50-million-year gap in geological time, there was a layer of coprolite nodules. Henslow shipped 2 tons of Felixstowe coprolites to Lawes to assay their phosphate concentration. Lawes reported that more than half the material was calcium phosphate.

In summer 1845 Henslow delivered a paper to the British Association for the Advancement of Science (meeting that year in Cambridge), to announce his discovery that coprolites at the base of the Red Crag in Suffolk were 'sufficiently abundant to make it worthwhile to collect'.

Henslow's talk launched a coprolite 'gold rush'. When a farmer produced a pile of coprolites dug from his land Henslow told him: 'You have found a treasure—not a gold-mine, indeed, but a "food-mine". Only find enough of them, and you will increase immensely the food supply of England and perhaps make her independent of foreign phosphates in case of war.'[5] As the 'coprolite rush' got under way, fields were excavated to a depth of 12 metres to access the base of the Crag deposits. One miner estimated he could harvest 60 tons of nodules each week.

You do not have to travel far outside Britain to find Miocene age marine sediments.[6] Through the second half of the nineteenth century, deep foundations were dug for fortresses to protect the city of Antwerp. These excavations revealed a sandy deposit with characteristic Miocene age fossils.

The Suffolk coprolite nodules had formed on an oxygen-starved ('anaerobic') sea floor over millions of years through the Miocene. They were smooth and brown, 2–10 centimetres across. The currents and tides winnowed away the sand and smaller shells. The coprolite nodules were accompanied by fossils: the teeth of the giant shark Megalodon, chunks of mastodon bones (an early elephant), skulls and bones of rhinos, crocodiles, beavers, and pigs which had been washed into the sea, along with the bones of walruses, dolphins, and whales. Mixed in were pebbles of far-travelled rocks, transported in floating tree roots.[7] And last, but most importantly for our story, there were 'boxstones': brownish cobbles of sandstone, in which the grains had become cemented together because the rock contained a fossil. Around 20 per cent included shells, whereas others had fragments of teeth or bone. As the thick shell or bone dissolved away, the calcium carbonate had crystallized around the sand grains to create the cobble. These were Miocene marine fossils.

By 1850 the basal Red Crag coprolite diggings had expanded across the estuary of the River Deben into Bawdsey and farther north along the coast to Boyton. Landowners were offered deals by which they would earn a share of the money raised by digging on their property. With yields of up to 1,200 tons, landowners could earn £2,500 per acre. A new quay was constructed at Stonner Point to take the shipments. By 1854, 12,000 tons of Suffolk coprolites were being raised each year.

To identify where to find the seam the foreman would arrive with a giant corkscrew, known as the dipper. The foreman screwed the dipper into the earth to see if it intersected hard rocks and nodules—the sure sign they had located the basal layer. As having their gardens dug up was the last thing the local cottagers would want, especially those only renting, they might offer a sixpenny 'tip' to encourage the foreman to prospect elsewhere.

Fossils, boxstones, rocks, and coprolites were all mixed together, but the super-phosphate factory would not accept the material until all the 'contaminants' had been removed. As the rocks passed in front of them, a crew of sorters, women and boys, turned the fragments, using a wooden scraper to avoid cutting their fingers. They 'worked at extraordinary speed... only hesitating when a fossil worth keeping for the collectors was spotted'. To a small boy, the fossil that could be sold for two pennies would mean a great deal, whereas a megalodon shark's tooth as large as a man's hand might raise a guinea.[8]

There was a ready market for the fossils, especially the more complete specimens, the larger shark's teeth or the shell of a turtle, and many geologists and museum curators came to call. Henslow himself acquired a great collection of fossils, including 400 specimens of bones from the ears of whales.

The diggings in Suffolk continued for more than a decade. However, prices fell when older beds of rich phosphate nodules were discovered in Cambridgeshire. The industry struggled on through the 1870s, but by the 1880s was facing competition from superior phosphate deposits in Peru and the South Pacific. Yet the coprolite business had brought money into these east Suffolk communities, funding schools, renovating churches, and improving the highways. Henslow's ambition had been realized.

Our challenge was to discover a British Miocene stone building. How are we doing? As neither coprolite nor fossil, the boxstones encountered in the Suffolk phosphate diggings were valueless rubble. Although we have failed to find a single Miocene age stone building, rubble-waste boxstones were used in the 1854–1860 restoration of the walls of Sutton and Shottisham churches in the Bawdsey Peninsula, as well as in some barns on the Sutton Hall estate.[9] In terms of our quest to find buildings of every geological age, this will have to do.

If 'British Miocene' was a collectable brand, with value determined by scarcity, boxstones would be as precious as topaz or emeralds. In case the value appreciates, a limited supply of Miocene boxstones can still be found emerging from cliffs at the base of the Red Crag formation, on the beaches at East Bawdsey[10] and Walton on the Naze.

*

The Miocene tectonic story across Europe is full of action.

At the beginning of the Cenozoic a mighty ocean, named Tethys (after a Greek sea goddess), lay between Africa and Eurasia, widening to the east. Tethys oceanic crust has since been subducted, drawn back into the mantle, and micro-continents including India and Arabia have collided with Asia as the intervening ocean closed. Today remnants of Tethys ocean crust are thought to underlie the Caspian and Black seas and floor the south-eastern Mediterranean.

Through the Miocene in the western Mediterranean ancient 'Tethys' ocean crust was subducted down to the north-west. At the same time the zone of subduction migrated south-east and a series of continental fragments, formerly attached to Spain and southern France, broke away, as rifting and new ocean crust formed behind them. In lockstep, Corsica and Sardinia rotated counter-clockwise away from southern France while the Balearic islands Mallorca, Minorca, and Ibiza pulled away clockwise from the east coast of Spain, leaving new sea floor in their wake. (If tectonics was national destiny, Corsica and Sardinia would be united and independent and the Balearics would also be their own country.) Another micro-continent became glued onto the Algerian coast of Africa while other continental scraps became Sicily and southern Calabria. A new westerly dipping subduction zone opened to the east of the Italian peninsula[11] and as it consumed old ocean crust brought new rifting and sea-floor spreading in the Tyrrhenian Sea to the north of Sicily, an area of volcanic activity today.

At the end of the Miocene, the Straits of Gibraltar closed tight and the Mediterranean was cut off from the Atlantic. In the hot desert conditions, water evaporated faster than could be replenished by the rivers. Sea levels fell 1,400 metres. In fact, so much water was lost from the Mediterranean that sea levels across the globe rose by up to 10 metres. The River Rhône thundered down a staircase of huge waterfalls, to reach the shrunken Mediterranean salt sea. Every few thousand years the barrier at Gibraltar broke, the Atlantic poured in through the most spectacular cascades, and within a few weeks the Mediterranean was full again.

Across Austria, to the east of the Alps, a great NE-trending fault was moving with sinistral motion. Vienna is situated on the western edge of a stepover ('pull-apart') hole in the crust, 200 kilometres long and 55 kilometres wide, that filled up with 6,000 metres of sediments from the eroding Alps. Budapest is a city of hot springs fed from volcanic activity that developed as the underlying mantle rose up to fill this enormous hole, melting as it ascended.

For a long time the rise of the Alps had been isolated from north-west Europe. In a 50-million-year 'pile-up', thick stacks of sediments (known as nappes—the French for tablecloth) peeled off the northerly subduction of the Tethys ocean crust. One kilometre-thick sequence of sediments piled up on top of another, leapfrogging nappe on nappe[12] and raising great mountains. The culmination came towards the end of the Miocene when the advancing Italian 'Apulian' plate finally collided to the north-west into the Jura mountains, whereafter,[13] as described by the British tectonician John Dewey, north-western Europe was 'caught in the jaws of a (tectonic) vice' (Fig. 40).

*

The tectonic vice enclosing Britain had the north-west corner of the Alps on one side, where the Italian promontory of the Africa plate was directing its force towards the north-west.

From the end of the Miocene, continuing into the Pliocene, the north-west compression led to tight folding above buried reverse faults in the Jura mountains.[14] The Jura are a product of the most recent Alpine collision, the crumpled consequence of a force directed to the north-west.

To the west of the Jura, on the edge of the Bresse Graben, lies Burgundy. Through reversing a rift-bounding normal fault, the tectonics raised a NNE-trending scarp of limestones and shales, with sun-drenched SE-facing slopes. The town of Dijon lies at the foot of a fold formed by an underlying reverse fault. Oligocene lake sediments outcrop at the bottom of the hill, while the scarp exposes Jurassic limestone and clays. The terroir of the great wines of Burgundy is a mix of geology and Miocene tectonics. Côtes de Nuit is the northern extension of the scarp, whereas Côtes de Beaune extends south. The Côtes d'Or ridge is almost 60 kilometres long, likely to have been raised by earthquakes close to magnitude 7.[15]

Since it formed 40 million years ago, the Rhine Graben has acted as western Europe's 'stress barometer'. Whatever the tectonics, the sensitive Rhine Graben faults respond. Beginning in the early Miocene, NW-directed compression crushed the southern half of the Rhine Graben, as far north as Baden Baden. The walls bounding the graben closed in, squeezing and raising the interior sediment pile by up to 1,500 metres. The graben floor was then eroded by the River Rhine, today down to elevations below 200 metres.[16]

The tectonic vice focused concentrated horizontal force, enough to raise the crystalline basement of the graben flanks into mountains: the 1,500-metre gneissic Black Forest mountains to the east and the

1,400-metre granitoid Vosges mountains to the west. The Jura mountains themselves reach more than 1,700 metres.

The 'blade' of the vice, pressing to the north-west, was only 200 kilometres wide. The resulting force-field had sharp boundaries. The Rhine Graben north of Baden Baden lay beyond the influence of the collision and graben margin faults were reactivated as strike-slip, creating local basins filled with Miocene and Pliocene shales.

★

The other side of the Miocene tectonic vice lay to the west of the Shetlands. The Norwegian Sea spreading ridge, which had been jammed by oblique collision on a prominent transform fault,[17] had finally become reconfigured and unlocked (see Fig. 46). Spreading was picking up speed, pressing strongly on the ocean margin to the south-east. Iceland had begun to form above a resurgent mantle plume, drawing the plates apart with the elevated lava piles pushing back into the plate to the east. A land bridge connected from an ice-free Greenland through to the emergent Iceland, along which plants like cornelian cherries and American white and red oak migrated through to Europe.

Spreading was now faster than the plates were separating (see Fig. 45). The continental crust on the ocean margin to the west of the Shetlands and around the Faroe Islands was being forced to shorten, through reverse faults and their overlying folds.

To see fully what was happening we would need to drain the sea to admire the new Miocene topography. New hills, new ridges, and new mountain ranges have been rising.[18]

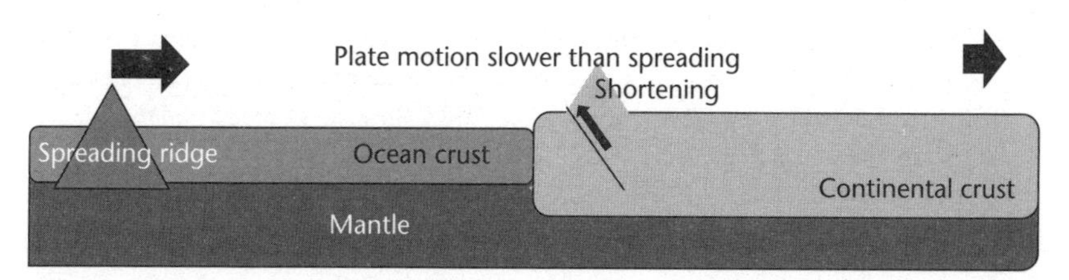

Figure 45. When the plate is moving slower than the opening at the spreading ridge we get regional compressional tectonics

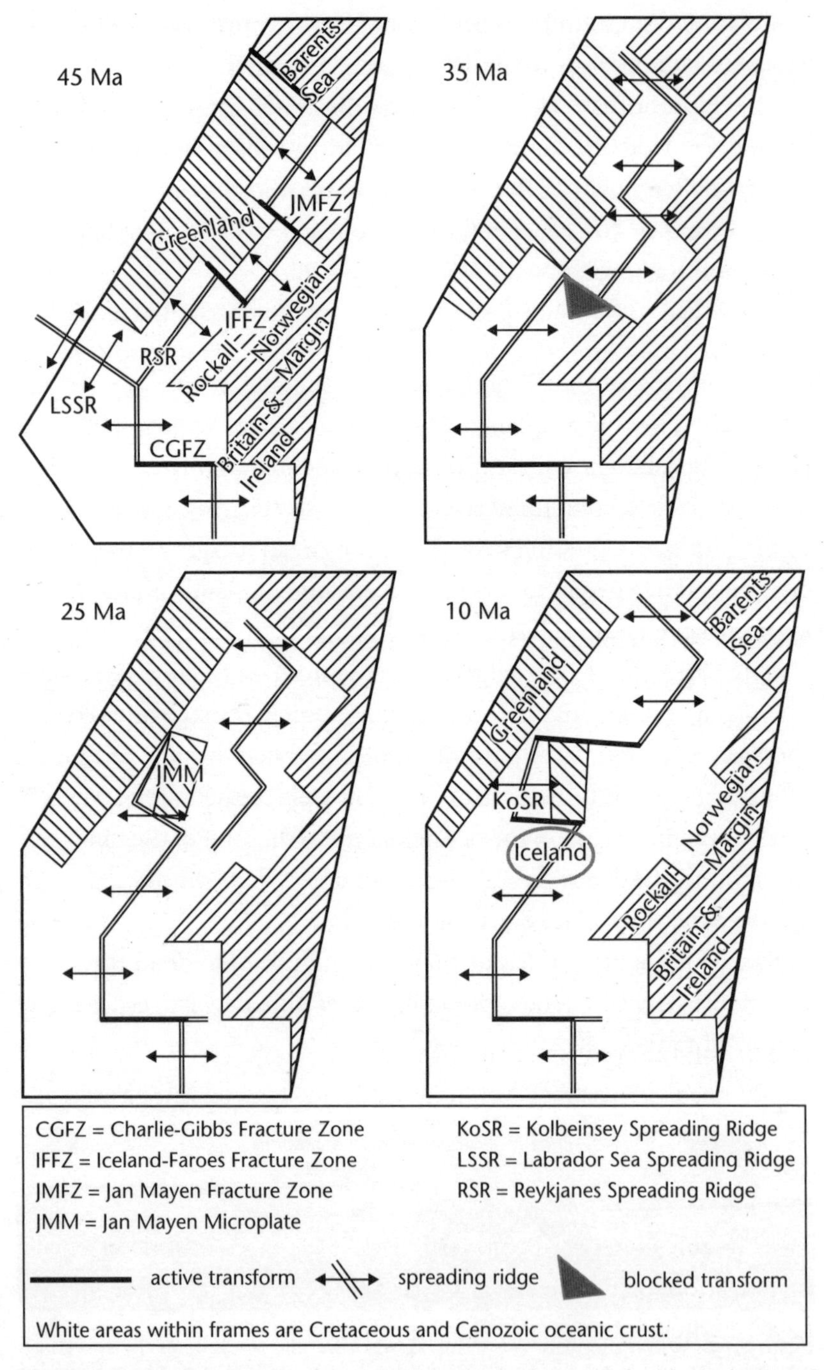

Figure 46. The resolution of the rotation of spreading and the blocked West of Shetland transform fault in the opening of the Norwegian Sea

Two hundred kilometres to the east of the Faroe Islands the NE-oriented Fugloy Ridge, made out of lavas erupted as the ocean opened, has risen more than 2,000 metres, through the mid to late Miocene. Follow the ridge to the south-west and the mountains pass into the East Faroe High. To the south of the Faroes the Faroes Bank Basin was being squeezed and raised.[19] In fact, the Faroe Islands owe their very existence to Miocene tectonics.[20]

In the Faroe–Shetland Basin the principal folded ranges, underlain by reverse faults, run NE–SW. The longest continues for 70 kilometres, raised by a formidable 3,000 metres.[21] The Faroe–Shetland escarpment that marks the edge of the sea-floor lava outcrop has itself been reactivated and raised 600 metres.

In the far north of the North Sea, the whole region rose up above sea level in the middle of the Miocene,[22] and there was a gap of about 10 million years without sedimentation.[23] North of 60°N, Lower Miocene sediments only survived in the centre of the basin.[24] In the middle of the Pliocene, when the vice was finally removed, the northern North Sea floor began to sink once again, accumulating a thick pile of sediments.

On one side, the Italian promontory colliding into the Jura mountains and the southern Rhine Graben, pressing hard to the north-west; on the other, the accelerating Norwegian Sea spreading ridge transmitting its forces to the south-east (Fig. 47). It was a tectonic collision of Titans. In the heart of the vice lay Britain, the land of 'Miocene geological amnesia'. Between the Faroes and the western Alps, in the middle of this corridor of powerful horizontal forces, something would need to give way, would need to buckle and crumple, like the hulk of a car in a wrecker's crusher.

⋆

True to the idea of a geological 'dark age', there is little scientific investigation of tectonics across Britain through the Miocene. We have hundreds of research papers on the British Paleocene volcanoes, and almost as many on the 'Pyrenean' faults and folds of the Eocene. The number falls off through the tectonics of the Oligocene, until by

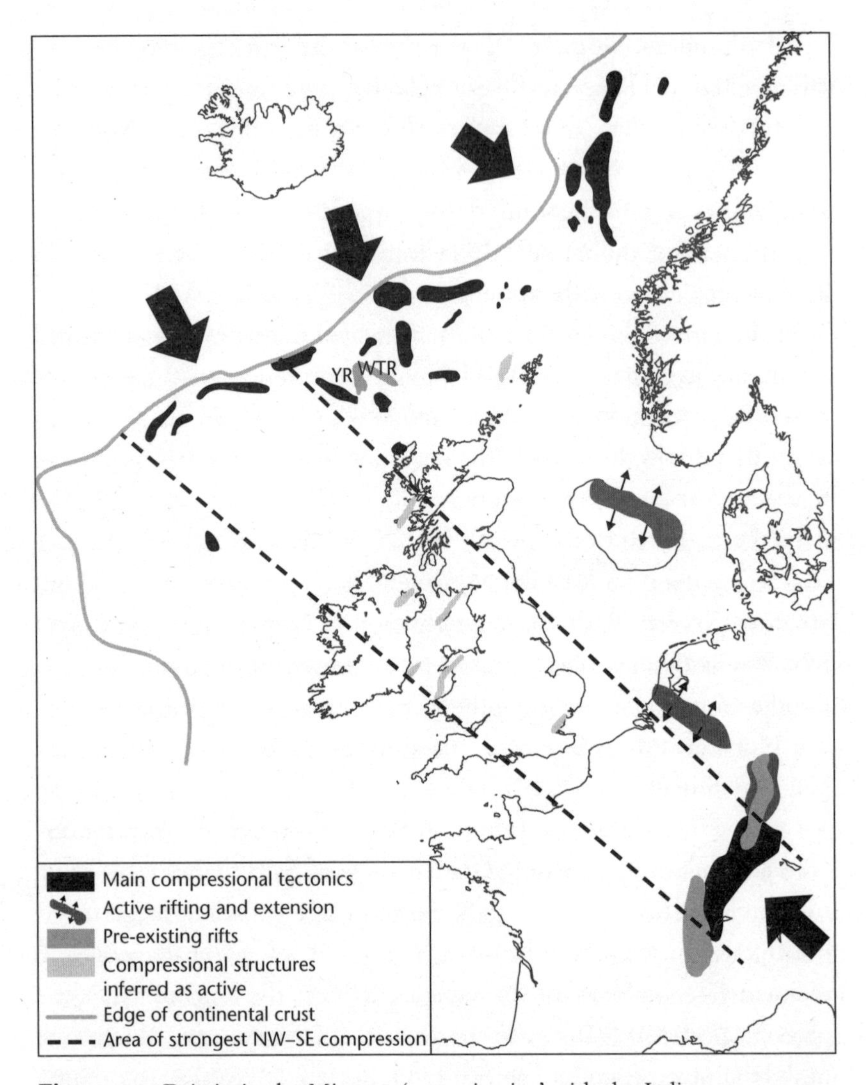

Figure 47. Britain in the Miocene 'tectonic vice' with the Italian promontory crashing into western Europe and raising the Jura Mountains to the south-east allied with compressional tectonics along the continent ocean boundary to the north-west. WTR = Wyville Thomson Ridge, YR = Ymir Ridge (active through Oligocene)

the Miocene I can only think of two or three research papers covering on-land British tectonics (there are more reporting on the offshore domain). Perhaps the best research on the topic is to be found in a thesis written in French.

Call it a challenge or see it as an opportunity. In the absence of Miocene sediments on land there is no record except the mountains and hills themselves (and maybe any veins in the rock opened by faulting[25]). Offshore, it is back to interpreting those seismic reflection profiles.

At the start of the Miocene the weakest links in the crust of the British landmass were the deep tectonic 'Celtic basins', stretched open and filled with sediment through the late Oligocene. In any collision, these basins were likely to be the initial shock absorbers. What would be the signs?

We should look out for upwarps: anticlines and monoclines above blind (and surfacing) thrust faults. Along the ocean margin to the west of the Shetlands the squeezing was directed towards the south-east, activating reverse faults oriented NE–SW.

In NW–SE cross sections across Rum and Skye, the south-east side of the Canna Basin is raised and folded. A borehole into the Oligocene basin encountered strata dipping 25–30 degrees down to the north-west.[26] This upwarp continues to the south, parallel to the Camasunary Fault, and continues through the islands of Coll and Tiree, on which all the lavas have been eroded, exposing the ancient gneisses of the Lewisian Shield.

To the north-east of the central volcano of Skye a spectacular ridge of basalt lava mountains runs for 30 kilometres along the Trotternish Peninsula, to the east of the Screapadal Fault,[27] rising to 713 metres at Storr. The Old Man of Storr pinnacle stands tall beneath the summit on the precipitous eastern flank. Under the lavas there is a thick pile of Jurassic sediments, intruded by great tabular sills.

Compressional uplift could explain the raised base of the lavas above the westerly dipping Jurassic sediments along the spine of the Trotternish Peninsula[28] (Fig. 48). The lava base slopes to the west from a maximum height of 600 metres, tilting from the Storr escarpment to sea level in 7 kilometres.[29] Although you might expect the pile to be sliding downhill to the west, deep glacial erosion in the Sound of Raasay (creating the deepest sea around Britain—a favourite haunt of

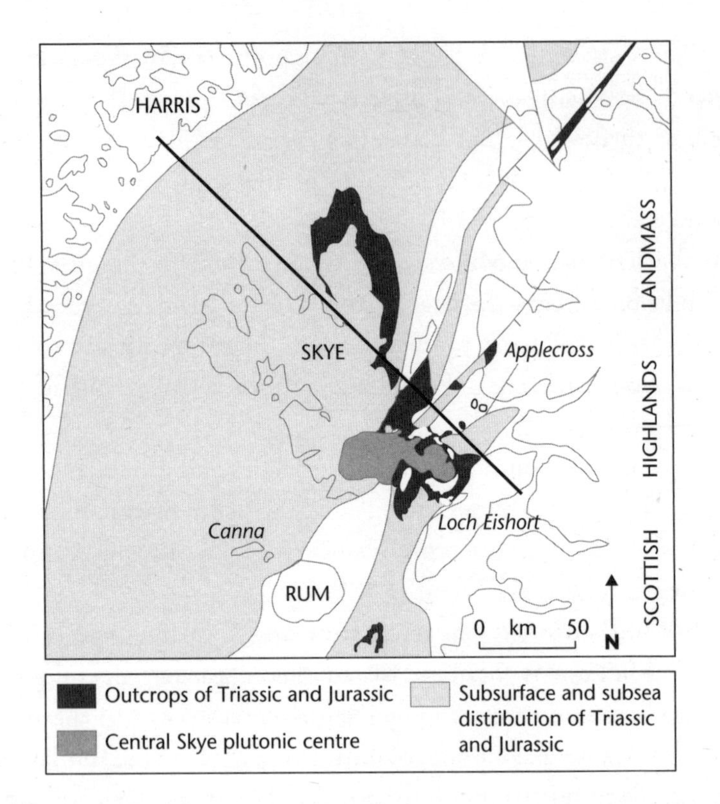

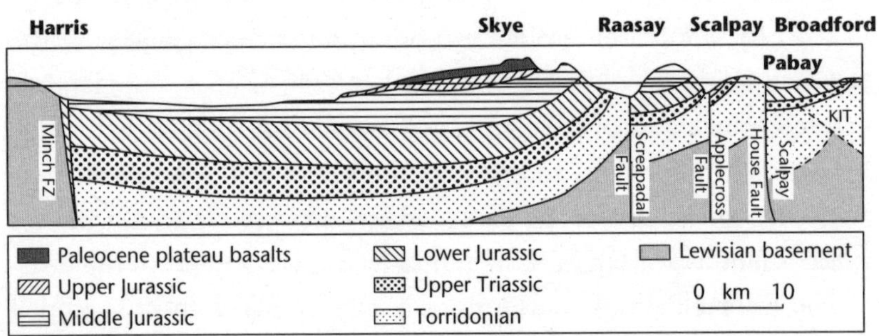

Figure 48. Probable Miocene uplift in north-east Skye

killer whales and submarines) means the great escarpment is tumbling towards the east.

This landscape seems chiefly the product of Miocene tectonics. Towards the north-eastern part of the peninsula lies the Quiraing terrain. Covering 8.5 square kilometres, this is not only the largest

Figure 49. Quiraing Landslide, north-east Skye

landslide in Britain but has even been declared by a foremost landslip expert to be 'the most beautiful landslide on earth'.[30] The ground comprises breaking waves, suspended in their motion, as apartment-building-sized blocks of lavas slump and topple, surfing on layers of clay and faults in the underlying Jurassic sediments, creating great cliffs and sculptural pinnacles (Fig. 49). There are the austere walls of the 'Prison', the 35-metre-high 'Needle', and the curious, hidden, and naturally lawned 'Table'. Some of the rotated blocks have left a little lake (a 'lochan') in their land-slipped wake. This terrain has featured as an other-worldly backdrop to movies, including the 2017 *King Arthur: Legend of the Sword*.

The Canna Basin is folded along a NNE–SSW trend that comes onshore in the north-western tip of Skye.[31] This, we infer, is Miocene tectonics when the original basin margin faults have also been uplifted and lost. The Skye Volcano has sustained the longest tectonic influence of all the British volcanoes.

A NE–SW ridge, mapped as a south-east continuation of the Southern Uplands Fault, raises the base of the Paleocene lavas in Northern Ireland from 900 metres below sea level in the Lough Neagh basin to more than 300 metres above sea level through Belfast.[32]

We believe the land was generally flat before the first Antrim lava flow arrived. The Lough Neagh basin was formed in the Oligocene but the upwarp of the south-east side of the Antrim Plateau looks to be Miocene.[33] The Lough Neagh basin switched from draining south to north as it was raised and eroded more than a kilometre through the Miocene.[34]

The largest of the Celtic basins, the St George's Channel Basin,[35] has been crushed and uplifted, principally on its south-east side along the Bala Fault. Along the western edge of St George's Basin, offshore from the south-east coast of Ireland, the basin has been exhumed more than 2 kilometres,[36] whereas the Mochras borehole core revealed 1,500 metres of uplift in the Miocene.[37] The uplifted and eroded basin margin has fostered a steep mountain torrent that has latterly led to river capture at Devil's Bridge in west Wales.[38] Oligocene lake sediments are found more than 100 metres above sea level at Trefynnon, to the north-east of St David's.[39] On the west side of the basin, in south-east Ireland, the topography reveals geologically 'recent' tectonics.[40]

On the northern edge of the Cardigan Bay Basin, the Eocene and Oligocene sediments rise up along St Tudwal's Arch, a NE-trending anticline. This marks the south-west end of the most spectacular 'Miocene' tectonics in Britain.[41] It runs from St Tudwal's Arch to the north-east, through the former pilgrimage destination at Bardsey Island, before ascending along 50 kilometres of the Llŷn Peninsula where the arch transforms into the Snowdonia mountain front (Fig. 50a and 50b).

The first person to suggest that the Snowdonia massif had risen through tectonic activity in the Miocene was Edward Greenly. After working for the geological survey in Scotland, Greenly spent the years 1894–1910, self-funded, and in partnership with his wife Annie, creating the most meticulous geological map of Anglesey.

On a clear day Greenly would turn and admire the great mountain wall of north Wales beyond the Menai Straits, with the summits rising in a regular arc to culminate in the centre at Snowdon (Yr Wyddfa). Recall the Paleocene igneous dykes that run across Anglesey and can

Figure 50a. Beaumaris Castle in the eastern corner of Anglesey looking south-east across the Menai Straits towards the 'Snowdonia Front' mountain 'wall'

still be found outcropping at elevations above 750 metres in Snowdonia (see Chapter 4). In the Isle of Man, far closer to the Irish volcanoes, the Paleocene dykes are rarely found above 30 metres elevation and never reach above 210 metres. After considering the question for forty years, in 1937 Greenly published a paper noting that the contrast in elevation between Anglesey and Snowdonia could not have existed at the time the dykes were intruded, because the magma would have poured out over Anglesey rather than reached elevations of 800 metres in the mountains. The topography has arrived late in the tectonic history, he declared: in the Miocene.[42]

The more recent champion of a Miocene tectonic explanation for the mountains of Wales is a distinguished French geomorphologist: Yvonne Battiau-Queney. In the 1970s she completed her PhD on the morphology of Wales, devoting eighteen pages to '*Le Front NW de la Snowdonia*'.

There is nowhere else like this around the Irish Sea—a low-lying, heavily eroded surface of ancient hard-wearing rocks, like Anglesey,

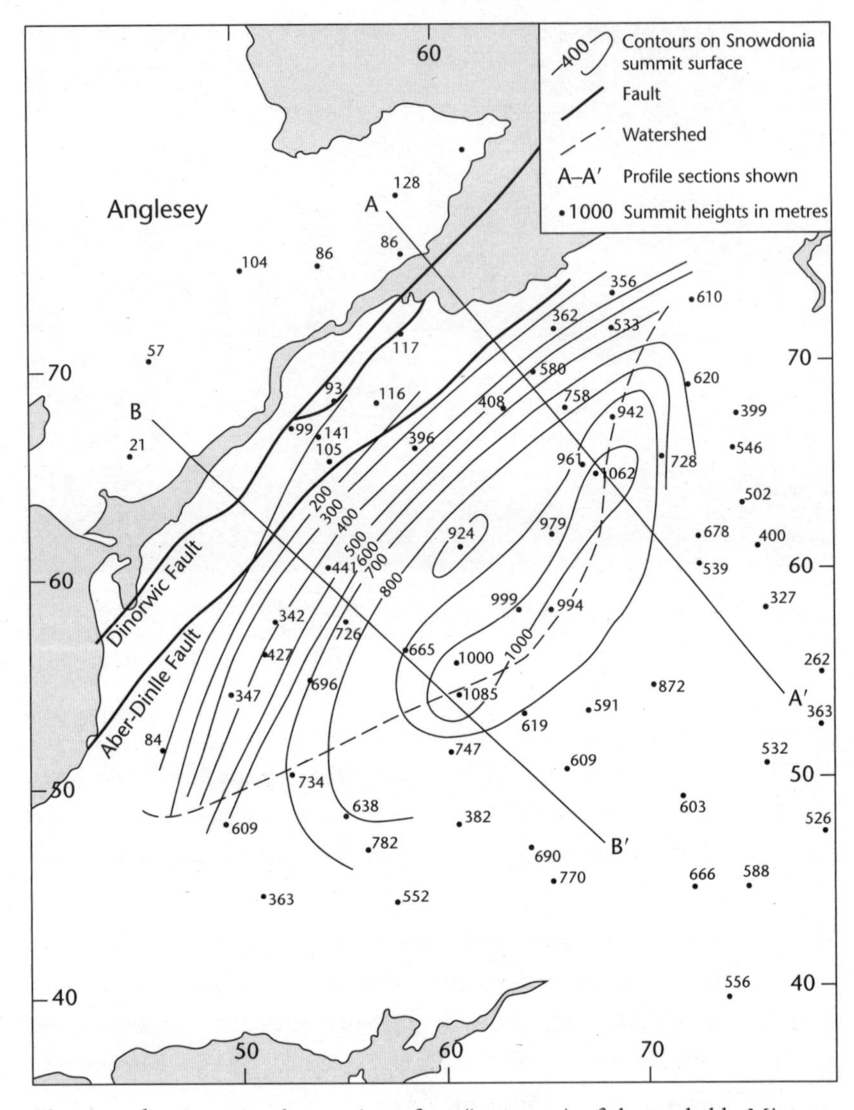

Figure 50b. Contoured summit surface (in metres) of the probable Miocene compressional uplift at the 'Snowdonia Front'

and its continuation into the Arfon marginal plain of north Wales, with the towns of Caernarfon and Bangor up against a linear mountain wall. The difference in heights between the Anglesey landmass and the summits of Snowdonia implies an intervening south-easterly

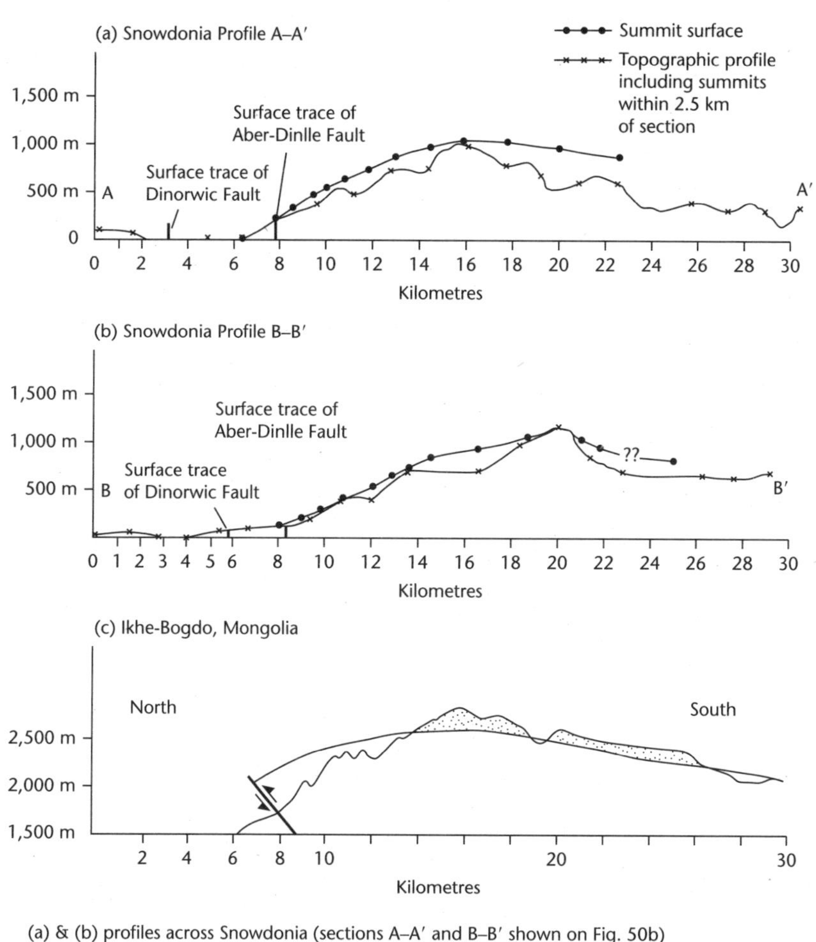

(a) & (b) profiles across Snowdonia (sections A–A′ and B–B′ shown on Fig. 50b)
(c) profile across the eastern Ikhe-Bogdo Mountains, Mongolia. Shows flexuring of Tertiary trachybasalts above reverse fault, last active in 1957

Figure 50c. Comparison between cross section through the summit surface of an active compressional range-front in Mongolia and the 'Snowdonia Front'

dipping reverse fault (most likely the reactivated Aber Dinlle Fault) with a displacement of at least 800 metres. Geomagnetic surveying indicates the Aber Dinlle Fault offsets a prominent Paleocene dyke.[43]

Unknown to Greenly, the NE–SW orientation is the preferred one for a reverse fault movement in the jaws of the Miocene tectonic vice. The overall shape of the mountain summits resembles an anticline above an underlying reverse fault. According to the 'smile principle' derived from observing the crack in a wall, displacement on any fault tends to be larger in the middle (i.e. at Mt Snowdon, Yr Wyddfa), tailing off to either end, to the north-east and the south-west.

The best analogue may be with the tectonics of Mongolia (Fig. 50c). There, far from a plate boundary frontline, the mountains absorb some of the collision between India and Asia, through movement on underlying reverse faults. Take away the yurts and the yak and some of these mountain fronts resemble Snowdonia. The faults are still active and the earthquakes of Mongolia are big: four larger than magnitude 8 since 1900 and thirty above magnitude 7.

The earthquakes along the Aber Dinlle Fault on the Snowdonia Front in the Miocene could have been the largest experienced across the whole region since the start of the Cenozoic, some 66 million years ago.

The complex geology beneath the Irish Sea to the north-east of Anglesey has been exhaustively surveyed while prospecting for oil and gas. Seismic profiles reveal an episode of late-stage compressional tectonics across the basin. Although Paleocene uplift and erosion have removed sediments younger than 200 Ma, this north-westerly directed compressional tectonics is confidently dated as 'Miocene'. The largest Miocene structure in the Irish Sea is the NE–SW-trending Ramsay–Whitehaven Ridge that runs from the Cumbrian coast north of Maryport to the Isle of Man.[44] To the east this uplift passes into the Carlisle Basin, where 2,500 metres of sediment has been eroded,[45] with perhaps 40 per cent lost in the Miocene.[46] The compressional tectonics continues into the Isle of Man, where NE–SW-trending reverse faults are claimed to cut Paleocene dykes close to Port Mooar on the north-east coast.

Throughout the Irish Sea basin, on almost every principal fault, there is evidence of inferred Miocene tectonics. NE-trending faults are overlain by anticlines. Northerly trending faults display[47] 'flower structures' and 'popups'—structural flourishes reflecting localized uplift, overlying compressional jogs (the opposite of pull-apart basins) in strike-slip, horizontal displacement faults.

In fact there is so much evidence for later fault movements beneath the Irish Sea, Miocene tectonics must also have been rampant in the surrounds, reawakening the faults beneath the mountains of central and southern Wales, playing on some of the larger structures across the Welsh Marches, chattering through the Cheshire and Lancashire basins, running along the NW-trending coastal boundary of Cumbria, and reactivating some of the ancient plate boundary faults, such as the Southern Upland Fault that offsets the buried valleys of the rivers Nith and Doon by 60 metres to the south-east.[48]

The structure of the southernmost of the Sticklepath pull-aparts, the 'Bovey Basin', has always been controversial.[49] In 1968 two geologists published a paper claiming there had been 6 kilometres of '*sinistral*' displacement on the Sticklepath Fault in the mid-Cenozoic. How come all the offsets are to the right? Simple! There had already been 10 kilometres of *dextral* displacement on the fault before the Cenozoic.[50] This elaborate proposal was triggered by confusion around the structure of the Bovey Basin.

Unlike the Petrockstow Basin, the Bovey Basin does not have a classic pull-apart shape (Fig. 51). We know what the basin structure must once have looked like, to form at a right-hand jog on a dextral fault. But today what we see is a pull-apart basin that has been squashed and partly rotated. On the northern side, the original sedimentary pile has been squeezed upwards, with some of the oldest sediments uplifted and exposed on the edge of the basin. Along the south-west edge, Devonian age sandstones have been thrust over the soft pull-apart sediments for many tens of metres, and the basin sides have closed in to isolate the 'Decoy Basin'[51] in the south-east corner.

The Bovey Basin was crushed and contorted after it was originally formed as an extensional pull-apart through the Eocene and Oligocene. This squeezing, we can infer, happened in the Miocene.

The classic pull-apart at Petrockstow has also been squeezed late in its life, with both sidewalls thrust over the basin.[52]

Away from the Celtic basins and we are moving into the heart of what makes geology such a pleasure. Speculative deductive reasoning,

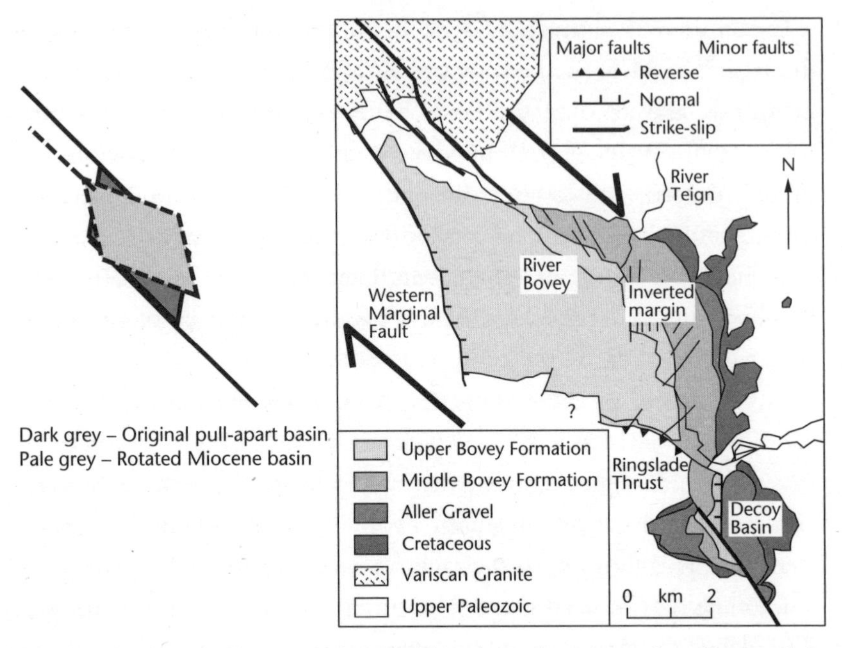

Figure 51. Bovey Basin on the Sticklepath Fault, squeezed and rotated in the Miocene

best argued in a pub over a pint of beer, without enough unambiguous evidence to settle the point.

Once you get into the swing of what to look for, you can 'label' a dozen candidate reverse faults.[53] 'Active in the Miocene?' 'Maybe.' 'I think it's your round.'

From Shrewsbury north-east towards Stoke-on-Trent, two Paleocene dykes (Butterton and Grinshill) outcrop, but neither can be traced farther towards their volcanic source, implying they have been raised along the NE–SW Red Rock Fault.

The band of Miocene compressional tectonics through western Britain runs NW–SE and is around 200 kilometres across, the same width as in the Jura mountains and the southern Rhine Graben, but only half the width of the continental-margin shortening found to the west of the Shetlands.

Farther to the east, into the central North Sea, we pass out of the direct influence of the tectonic vice: instead of uplift, Miocene sediments have accumulated in NW–SE-trending basins, stepping to the right[54] and trending away from Britain.

Squeeze the crust NW–SE and if there is no impediment it will expand NE–SW. We know that the common planar cracks (or 'joints') in rock exposures in southern Britain and north-east France run consistently NW–SE.[55] These joints are mementoes of the Miocene.

Perhaps the most intriguing candidate for Miocene tectonics is London's elusive 'Greenwich Fault'. First discovered in 1860s railway construction 3 kilometres south-east of Greenwich, the ENE-oriented 'fault' was soon projected as spanning the whole of Essex. It took another century before the speculative eastern extension was dropped. The structure is now confidently mapped for 25 kilometres under south-east London. It starts with the NE-trending, 25-metre-offset Wimbledon Fault beneath the town's former greyhound racing stadium. This passes into the Streatham Fault which at Balham is a monoclinal fold intersected by faults. This continues to the north-east into the Greenwich Fault, which is also principally a 'blind-thrust' monocline, running along the northern boundary of Greenwich Park

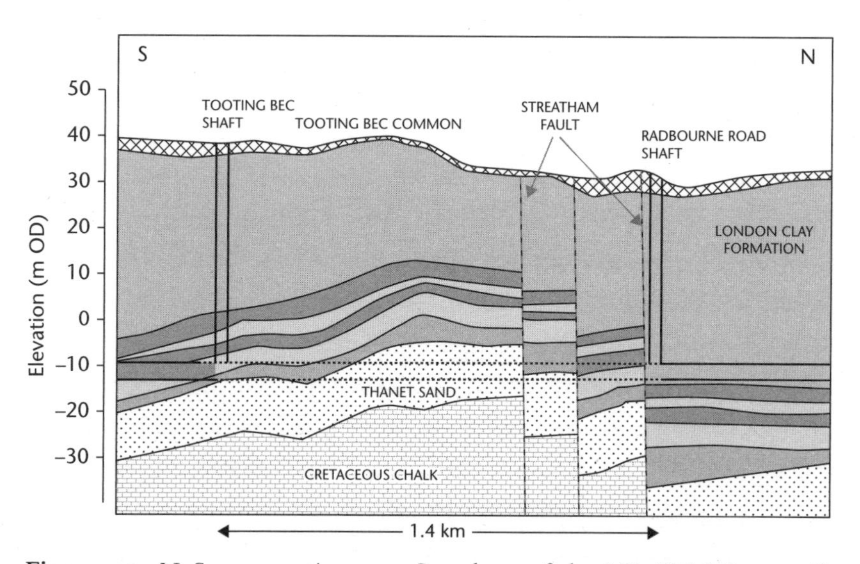

Figure 52. N–S cross section near Streatham of the NE–SW Miocene (?) Wimbledon – Streatham – Greenwich Fault in south-east London

and under the Royal Docks.[56] Displacing all the Eocene strata in its path, from its orientation, the faulting is dated Miocene[57] (Fig. 52).

A potential Miocene fault even underlies Windsor Castle. In the surrounding swampy flood plain (how do you think Slough, across the river, got its name?) the only prominence on which to locate a castle was the 30-metre-high Windsor chalk hill at a bend in the river, formed by a 2-kilometre-long, 150-metre-amplitude, NE–SW monocline.[58] The Royal Family have picked a prime tectonic location (probably a 'compressional jog' in a strike-slip fault zone), which not long ago, geologically speaking, was generating earthquakes.

*

'For how many years can a mountain exist, before it's washed to the sea?' is the title of a scientific paper on the persistence of topography (as well as a Bob Dylan lyric from 'Blowin in the Wind'[59]).

When the buildings have all been long destroyed, archaeologists excavate the 'midden'—the rubbish dump. The equivalent to the midden, for our story, is to study what was eroded from the land and has ended up deposited beneath the surrounding seas. The higher the mountains, the more sand and mud were transported by rivers and eroded off coasts to be deposited in water depths beyond the influence of storm waves.

Seismic reflection data, collected in the course of oil and gas exploration, can be used to measure how the amount of material worn off the landmass has varied through the past 60 million years.

One area continuously fed with sediment worn from Britain lies on the north-west continental margin to the west of the Shetlands.[60] Sediment production from erosion responds to uplift, which on this evidence has been sustained and even accelerated through the past 50 million years. An estimated 80 per cent of all the sediment volume arrived long after the intense volcanic activity and uplift of the Paleocene, with a mid-Miocene peak between 15 and 10 Ma. The mountains did not exist for long before they got washed to the sea.

9

Dambusters

In the 1830 first volume of his *Principles of Geology*, inspired by Robert Hooke's theory that earthquakes caused land uplift, Charles Lyell scratched around for well-observed examples. The evidence was scant. There were reports of a great landscape step, named by the locals 'Allah Bund' (God's embankment), that had suddenly burst into existence in a great earthquake in the desolate 'Rann of Cutch', on the eastern side of the Indus delta in 1819. And there was a letter from the writer, artist, and naturalist, Mrs Maria Graham from Valparaíso, Chile, reporting that the Chilean coast had risen by 3–4 feet after a great earthquake in 1822. As a lawyer, Lyell knew this was not enough to confirm his core belief that Earth history could all be explained by the processes we see today.

When Darwin left on the voyage of the *Beagle* on 27 December 1831, in his cabin he had a prized copy of Lyell's first volume, including the well-thumbed chapter on earthquakes and uplift, which lauded Robert Hooke. In the late morning on 20 February 1835, while resting on the ground in a wood on the edge of the settlement of Valdivia, on the southern coast of Chile, Darwin had the amazing good fortune to experience, at a harmless distance, a great earthquake. After sailing farther up the coast to the devastated city of Concepción, Darwin had set off on a lengthy trek over the Andes before Captain Robert Fitzroy surveyed 3 metres of accompanying coastal land uplift on the nearby Isla de Santa María. By the end of 1835 Lyell had received an account, sent by Darwin, describing the coastal uplift that had accompanied the great Chilean earthquake earlier that year. This account was given priority in the second edition of Lyell's *Principles of Geology*.

Yet even before writing his book, Lyell had heard stories from Sweden which appeared to confound the theory that uplift accompanied earthquakes. Around the coast of the Gulf of Bothnia, in the heart of the Baltic Sea, either the sea was retreating, or the land was rising, but without seismicity.

The stories went back centuries. In 1491 four townsmen from the port of Östhammar, north of Stockholm, made representation to the archbishop in Uppsala, the acting head of government, to demand their port should be relocated as its rocky harbour had become too shallow. In 1648 Queen Christina allowed the northern port of Luleå to relocate to the east to keep up with the retreating shoreline. A survey at the start of the eighteenth century found that older inhabitants shared stories on how their sea was becoming shallower, how the water appeared to be draining away. These were rocky boulder-strewn coasts, so siltation was not an explanation.

The city of Stockholm even owed its existence to falling sea levels. Viking sailing boats on Sweden's third largest lake, Mälaren, could access the Baltic Sea through a narrow channel. Over the centuries the level of the lake became generally higher than the sea. Since the year 1200 this narrow channel had turned into a rapid, so a boat had to unload and transfer its cargo overland to another boat which could continue the journey. And so was born a settlement that provided carriers, and places for sailors to eat, drink, and sleep: Stockholm, the city born from receding sea levels.

In 1743 the Swedish physicist Anders Celsius set out to estimate how fast water levels were changing. He found an ancient seal-hunter's platform, known to have been carved at sea level towards the end of the sixteenth century, and measured its current height out of the water.

Celsius also had his own water-level mark carved on a former seal rock in the south-western part of the Gulf of Bothnia (Fig. 53), so that in the future the process could be closely monitored. Within the following decades another eight sea-level markers were cut and clearly dated so that future generations could measure the change.[1]

Figure 53. Celsius sea-level mark boulder (with later additions) in the Gulf of Bothnia, Sweden

In the summer of 1834 Lyell travelled to Sweden to investigate the evidence for himself.[2] He measured the height of the mark cut by Celsius and calculated an 8-millimetre annual retreat in sea level over the previous ninety years.

In Stockholm he estimated sea level was falling at 5 millimetres each year, but he calculated twice this figure along the northern Gulf. Sea level appeared stable on the south coast of Sweden and the west coast of Norway, but at Oslo the land appeared to be rising at much the same rate as in Stockholm. This was not, as initially thought, some consequence of latitude, but fell away with distance from the Gulf of Bothnia. Lyell also mapped fossil beach ridge deposits between 1 metre and 27 metres elevation, containing the same shells as in the modern Baltic.

In 1835 Lyell wrote:

> In regard to the proposition, that the land in certain parts of Sweden is gradually rising, I have no hesitation in assenting to it after my visit.... I have no doubt that the rate of elevation is very different in different places.... I may be allowed to congratulate the scientific world that this wonderful phenomenon is every day exciting increased attention.

The account (sent by Darwin) describing the coastal uplift that had accompanied the great Chilean earthquake reinforced Lyell's original belief that uplift accompanied earthquakes. And yet in the Gulf of Bothnia, uplift was proceeding continuously and uneventfully. He could not explain this Swedish exceptionalism.

*

Following the Miocene hiatus, by the middle of the Pliocene, sediments once again began to accumulate in England. The British 'geological time recorder' had switched back on again. From Gedgrave to north of Aldeburgh in Suffolk, there is a 20-metre-high and 12-kilometre-long ridge of Pliocene 'Coralline Crag', rich in fossil (coral-like) bryozoans. The Coralline Crag is surrounded by later Red Crag sediments: a shelly, quartz-rich sand, coloured a rusty red, deposited in a warm shallow sea around 2.5 million years ago.[3] Then came the Norwich Crag formed between 2.4 and 1.8 million years ago, followed by the Wroxham Crag, containing sand and gravels deposited from a mighty River Bytham that drained the whole of central England.

Each formation reflects a climate colder than the one before. Next, it would not be a river delta overrunning northern Norfolk, but an ice sheet.

In 1837 the Swiss zoologist and expert on fossil fishes, Louis Agassiz, announced that during some geologically recent 'ice age' a much larger area of his country had been overlain by glaciers. He speculated that at the same period ice sheets had covered northern latitudes, including Scandinavia and northern North America.

The idea was explosive. Lyell and the leading geologists in England had successfully argued that the processes of the present could explain the past—that catastrophism was an unnecessary Continental philosophy. Yet, here was a catastrophe theory involving not fire or mountain building but frozen wastes of ice. In 1840 Lyell travelled with Agassiz around northern Britain, including to the Lyell family estate in Scotland, and returned a believer. A great ice sheet and glaciers could explain the landscapes of Scotland and northern England.[4] He wrote a detailed research paper with Agassiz supporting their case.

Yet after Agassiz had returned to Switzerland, Lyell came under pressure from Murchison, president of the Geological Society, backed up by other learned members, to recant this 'heresy'. His joint research papers with Agassiz remained unpublished. Lyell spent the rest of his life arguing against the theory of the Ice Ages. Instead, he attempted to explain all the glacial signs as products of floating icebergs, even though this required dramatic changes in the level of the land by hundreds of metres in the recent past—one might think, a far more catastrophic theory than that the mountains had been shrouded in ice. As we shall see, he was relying in this argument on observations collected in the Scottish Highlands by his friend and colleague Charles Darwin.

In 1865 the Aberdeen-based geologist Thomas Jamieson, far from the Ice-Age-denying Geological Society,[5] made the connection between the ice sheets and the uplift:

> It is worthy of remark that in Scandinavia and North America, as well as in Scotland, we have evidence of a depression of the land following close upon the presence of the great ice-covering.... It has occurred to me that the enormous weight of ice thrown upon the land may have had something to do with this depression.... If... the matter on which the solid crust of the earth reposes... is in a state of fusion, a depression might take place from a cause of this kind, and then the melting of the ice would account for the rising of the land, which seems to have followed upon the decrease of the glaciers.

This proposal was way ahead of its time.

In the late 1800s, geophysicists considered the interior of a cooling Earth to be as hard as the strongest steel. This contrasted with Darwin's experience of earthquake shaking and volcanic eruption in Chile, which had convinced him[6] that 'the molten matter beneath the Earth's crust possesses a high degree of fluidity, almost like the sea beneath the Polar ice'. At the start of the twentieth century, after the discovery of radioactive heating, the significance of those eighteenth-century Swedish water marks and Jamieson's proposal became more widely appreciated. Knowing the speed and timing of the slow return flow of the mantle, one could measure the tendency of the mantle to flow (its 'viscosity'). For the geologist Arthur Holmes, Swedish uplift revealed how mantle currents could power continental drift.

So how come the ice sheets merit their own chapter in a tectonic history?

The ice sheets were so vast and caused such enormous changes that they are worthy accomplices to tectonics in the business of landscaping. Ice sheets succeeded in transforming many of the river systems of central England and turned Britain back into an island. In thousands of years the ice sheets raised and lowered land at speeds a hundred times faster than those of the local tectonics. And as we shall see, the ice sheets also succeeded in triggering some very large earthquakes.

★

The fresh wind-blown snow on the high mountains drifted in gullies. Even in summer the snow did not melt and the first snows of the following autumn fell on the surviving snowpack. The Gulf Stream was far to the south of its current configuration and the temperatures across the North Atlantic had fallen by several degrees. As the snow accumulated it became compressed into ice. The mountain glaciers merged into an ice sheet like those of Greenland and Antarctica today, from which tongues of ice flowed down to lower elevations.

The past 2 million years of the Quaternary have seen repeated episodes when ice accumulated over the mountain uplands of northern Europe, and flowed out overland to the south, as well as to ice shelves on the edges of the nearby ocean.

At times the ice sheet from Scandinavia advanced across the North Sea into East Anglia, combining with the ice sheet that emerged from northern Britain. The farthest advances of the ice left great untidy piles of moraines: mud and stones scraped off the land and deposited by the melting ice in vast waste tips. Certain unique rocks, like boulders plucked off the Ailsa Craig microgranite, became 'calling cards' along the whole east coast of Ireland and into the coast of Lancashire, north Wales, and Pembrokeshire, revealing how the lowland ice sheet swirled in its path, like an abandoned fire hose.

South of the ice lay treeless tundra. At the culmination of the last glaciation the tree line had retreated to the latitude of the Pyrenees and Provence.

Everywhere the ice carved and polished the rocky outcrop. Then all the debris from the mountains got dumped in the lowlands. When the ice flow was moving quite fast, the debris was fashioned into a sculpted landscape of drumlins, like moguls on an overused ski-slope, but more often there was banal, formless glacial till.

You can't help hating the Ice Ages; they were like barbarian interludes in Earth history. The ice sheets were such vandals, such destroyers of the evidence of previous shorelines and rivers, that we can only reconstruct the chronology of the ice sheets from the sedimentary record preserved in the floor of the oceans.

Only where the mountains protruded above the ice, as 'nunataks', some Alpine-style serac peaks survived. James Boswell first noted the Cuillins on Skye were 'a prodigious range of mountains, capped with rocky pinnacles in a strange variety of shapes', resembling mountains in Corsica, unlike anything else in Britain.[7] To the west of the ice sheet's inland culmination, the ice never overwhelmed the remains of the volcano's former magma chamber.

The peaks of the last three glaciations in Britain have been dated to (i) between 478,000 and 424,000 years ago (the most extensive of them all, the 'Anglian'); (ii) from 191,000 to 130,000 years ago (the 'Wolstonian'); and (iii) from 29,000 to 14,000 years ago (the 'Devensian').[8] (Across western Europe the Anglian becomes the Elsterian, the Wolstonian becomes three phases of ice advance including the Saalian, and the Devensian becomes the Weichselian.)

The Anglian ice sheet had extended deep into the Netherlands and Germany and reached the edge of north London just south of Hornchurch. A tube ride on the High Barnet branch of the Northern Line would have been sufficient to see the stagnant edge of a great ice sheet, with the terminal moraine exposed in the railway cutting at the Finchley and Hendon stations.

At the climax of the most recent Devensian glaciation, around 20,000 years ago, the ice sheet covered all but the fringe of southern Ireland, almost all of Wales, then skirted the Peak District before reaching the coast of northern Norfolk (Fig. 54). Worldwide sea levels were 130 metres lower, exposing vast areas of what is today the continental shelf.

Beneath London, the chalk formation forms a syncline or shallow 'U' shape. The rain and snowmelt falling on the Chilterns (to the north) and the North Downs (to the south) replenishes an over-pressured 'artesian' aquifer, confined by impermeable clays. At the bottom of the 'U' in central London, water would burst out of any well or borehole cut into the buried chalk. Before industrial over-pumping in the nineteenth century, natural artesian pressure fed the Trafalgar Square fountains.

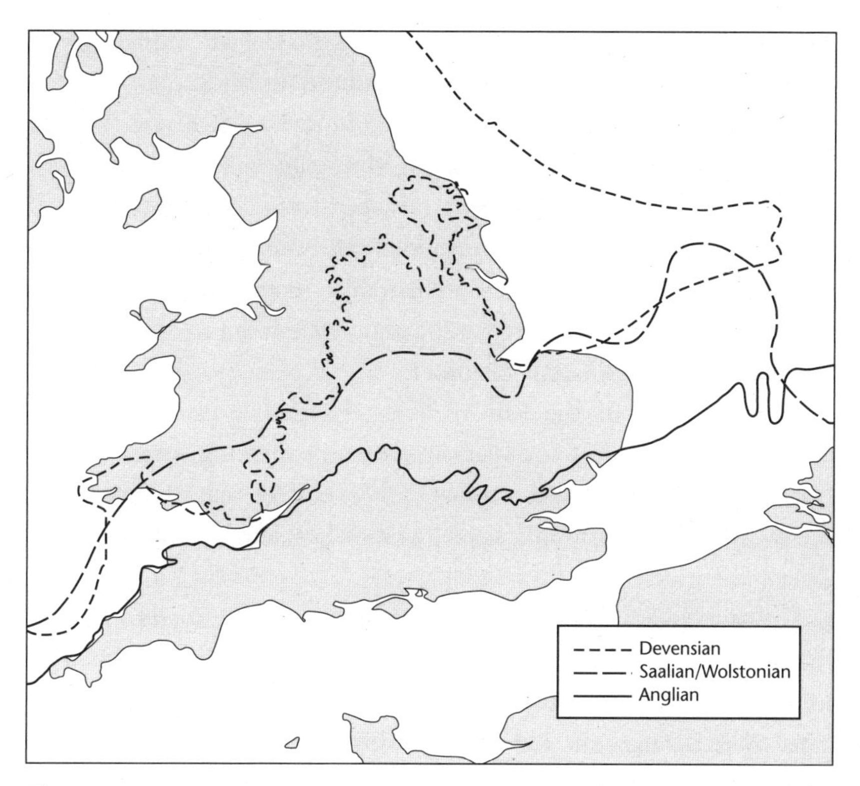

Figure 54. Maximum southerly extent of the ice-fronts of the principal Ice Age glaciations

Through the Ice Age, at a small number of locations across London, over-pressured water forced its way to the surface, feeding house-sized ice mounds known as pingoes. The underlying chasms have provided hazards to tunnellers and builders. One of the largest of these 'seething' hollows, close to the River Thames, was avoided for housing development and thereby became the site for the iconic Battersea Power Station.[9]

*

The landscape changes wrought by the ice were true 'terraforming'.

Around Britain, the main transformation concerned the wholesale diversion of rivers. The ice sheets advanced from the north and the north-east, damming any river that formerly flowed in those directions.

The water level would rise in an ice-dammed lake until it could overflow the lowest elevation in the surrounding landscape.

At 17,600 years BP (shorthand for 'Before Present', taken as the year 1950), in north Yorkshire, an ice sheet advancing across the North Sea blocked the flow from an easterly flowing river system in the coastal Vale of Pickering. The ice-dam ponded the water until it overtopped to the south-west, cutting the Kirkham Valley gorge. Today, a river that used to flow a few kilometres to the sea now follows a 160-kilometre diversion.[10]

Lake Humber in the Vale of York changed level several times between 17,000 and 15,500 years BP, each time creating beach terraces from 10 to 40 metres above today's sea level. The lake overflowed to the south at Gainsborough, north-west of Lincoln, but did not carve a permanent diversion.[11]

The current Welsh headwaters of the River Severn used to flow to the north, along the path of the River Dee around Chester. Then an ice sheet moved across the Irish Sea, advancing over the lower River Dee and creating Lake Lapworth.[12] The water levels rose until they overtopped the hills to the south. The overflow cut the Ironbridge Gorge, permanently diverting the river, so that the central Welsh mountains became the headwaters of the modern River Severn.

Another glacier dam arriving from the north-east blocked the path of a great river that drained the English Midlands into the North Sea. During the Anglian glaciation, vast Lake Harrison[13] first overtopped the landscape to the south-east, becoming the northern headwaters of the River Cherwell and Upper Thames. However, when even this outlet became blocked by ice, a new valley was eroded that allowed the River Severn to capture the modern Warwickshire Avon, which today flows into the Severn at Tewkesbury.

Before the Anglian glaciation advanced into the south Midlands there was a proto-Thames that flowed to the north-east through the Vale of St Albans, across Essex and Suffolk.[14] The ice sheet dammed this route, forcing the river to make two shifts to the south and find its current course through London.

A great lake in the Vale of Oxford, dammed from the North Sea, overflowed across the Chilterns to create the modern Goring Gap. Before this time the lower Thames River had only included streams flowing to the south of the Chilterns and the Berkshire Downs. Rapid erosion has meant that all the streams flowing into the Vale of Oxford have had to lower their base levels, creating strange incised topography, like the steep slopes below the White Horse at Uffington.

Even telling these stories is a challenge. What do we mean by a 'River Thames' or 'River Severn' whose course and catchments were completely different to those today? To paraphrase Heraclitus: through repeated glaciations, 'you never step in the same river course twice'.

The River Severn was the great 'winner' from the Ice Ages. For a time, the River Thames had gained a catchment to the north, and even into Wales, but then it lost it again to the Severn. During the last glaciation the River Severn carved a gorge from Gloucester down to the lowered sea level, forcing tributaries to carve similar gorges, including the Bristol Avon and the Lower Wye valley. The lowered base level eroded deep valleys on the western edge of the Cotswold Hills, producing great sheets of fan gravels full of Jurassic limestone pebbles.

★

By far the greatest example of glacial terraforming around Britain was the creation of the Dover Straits.[15] Before the Ice Age, the Wealden anticline continued uninterrupted into the Boulonnais coastal hills of France south of Calais. On the French coast we still have the 150-metre chalk cliffs next to Sangatte at Cap Blanc Nez (the continuation of the white cliffs at Dover). To the south there are Jurassic sandstones and clays at Cap Gris Nez, in the axis of the anticline and the nearest point to England.

The geologist Alec Smith was the first to propose that the opening of the English Channel had been catastrophic.[16] In the 1960s, in collaboration with Bristol University, Smith led a campaign of 'seismic reflection' research cruises to map what lay beneath the floor of the Channel.

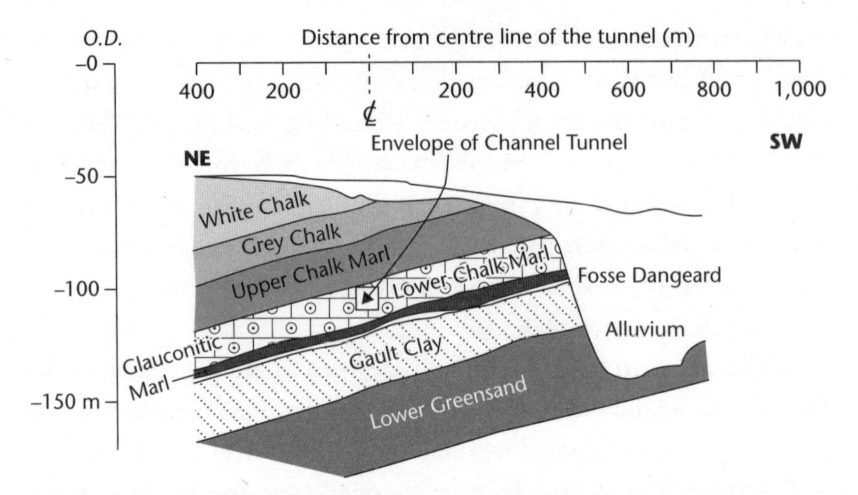

Figure 55. Proximity of the Channel Tunnel to the deepest Fosse Dangeard holes

The chief evidence of catastrophe was a series of huge holes and deeply scoured channels. The holes had first been discovered in the Straits of Dover by French seabed surveys in the 1920s. However, it was seismic reflection surveys undertaken to identify a route for the Channel Tunnel that showed the bedrock holes were far deeper, but largely infilled with later sediment (Fig. 55). Woe betide if tunnellers should accidentally encounter one of these cavernous holes! They were named after the French geologist Louis Dangeard (who had completed his PhD on the seabed geology of the Channel): the 'Fosse Dangeard'.

The holes measure a kilometre or more across; the seven deepest are incised down to 140 metres below sea level, burrowing deep into the Cretaceous bedrock. They are clustered in a 7-kilometre-wide WNW–ESE-oriented belt, parallel with the strike of the Cretaceous strata. The Channel Tunnel runs to the north of the Fosse Dangeard plunge pools, passing closest between kilometres 42 and 38. The holes cut deeper than the tunnel, which at its lowest plumbs 115 metres below sea level.

Only one phenomenon, Alec Smith argued, could produce these chasms: enormous waterfalls, acting like giant water drills.

What did this landscape look like when the catastrophe happened, 450,000 years ago? Did the chalk hills on the southern side of the anticline continue all the way from the Seven Sisters to the French coast at Étaples at the mouth of the River Canche? Or had the sea already broken into the weak sands and clays exposed in the heart of the anticline? We know the North Downs between Dover and Cap Blanc Nez were a continuous range of hills, but there were wind-gaps and stream valleys. The pattern of Fosse Dangeard holes suggests there may have been a line of southerly facing cliffs over which the water cascaded.

Yet to burst the barrier of hills that spanned from Kent to the Boulonnais would first require loading up a vast reservoir. Remember this was in the middle of a fierce Ice Age, when global sea levels were 100 metres lower than today. The ice dam spanned from Norfolk to the mouth of the Elbe, the ice sheet flowing from Scotland seamlessly merging with the great ice sheet sourced from the mountains of Scandinavia. With all the water flowing into the southern North Sea from the Rhine, the Meuse, the Elbe, the Weser, and the Thames, from Atlantic fronts, from the summertime thaw in the ice sheets, the level of the southern North Sea lake began to rise, year after year. First, to breach today's sea level. And still the water level rose, more slowly now that the lake covered a wider area (Fig. 56). It was full of icebergs, the larger blocks grounded in deeper water, farther offshore.

The water started to back up all the river estuaries that fed into the southern North Sea, inundating the surrounding flood plains. Sediments deposited from the great southern North Sea lake have been found in the cliffs of northern Norfolk and on the borders of Germany and Holland, up to 30 metres above sea level. The lake now covered at least 50,000 square kilometres, submerging all the Netherlands, encroaching on Duisburg, Germany (at 31 metres above today's sea level) where the River Ruhr joins the Rhine, and inundating southern Essex, northern Kent, and London as the Thames backed up to Maidenhead (at 32 metres).

Based on the height of the lake sediments, the overflow point on the chalk hills was no higher than 35 metres (above today's sea level).

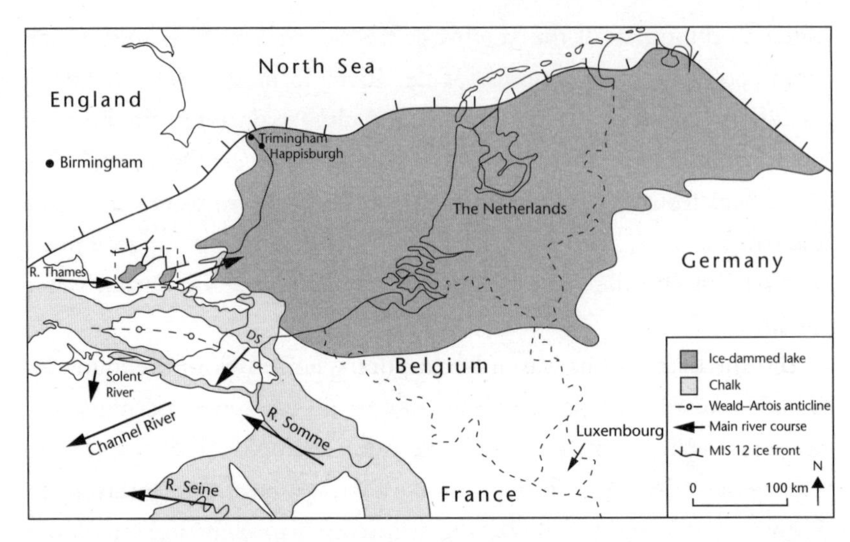

Figure 56. Extent of the 450,000 year BP ice-dammed lake in the south North Sea

What happened next happened fast. First a trickle, then a flood, as water began to flow over the chalk-ridge dam. Soon thousands of cubic metres of water were flowing each second through what, days before, had been an empty chalk valley with its arctic vegetation of lichens. The rock was soft and easily eroded. Every metre the lake level dropped liberated another 50 cubic kilometres of water, plunging over a precipice tens of metres high. The plunge pools beneath these waterfalls were drilled by the falling water down to 140 metres below sea level. This was the greatest volume of flood discharge on Earth over the last million years.

As one channel became blocked by icebergs, the power of falling water carved another. The noise must have been deafening. Over days the site of the waterfalls shifted as the underlying rock was eroded, carving fresh craters until a wide gorge had been carved through the chalk hills down to far below current sea level, filled with a wide, raging torrent.

The North Sea lake really was 'the Great Flood', a deluge of biblical proportions, reaching higher water levels than any flood or sea level since. And then, most remarkably, the flood drained away in days, like unplugging a bath.

Did any early hominid bear witness, having to shift their shoreline camp to accommodate the water rising a total of 130 metres, and then hearing the distant thundering waterfalls heralding the rapid recession?

Now, for the first time since the Pliocene, Britain was an island again.

But the island status was only temporary. For much of the following 400,000 years, with lower glacial sea levels, the landmass of Britain was once again a north-western peninsula of Europe. Island status was only renewed during the higher sea levels of the warm interglacials. However, whenever the climate was most amenable to settlement, the Channel moat seems to have prevented new arrivals.

At the end of the Anglian glacial, the Thames and Scheldt rivers continued flowing out through the 'Mid-Channel Gorge', whereas the Rhine returned to flowing north into the North Sea. As sea levels rose the gorge was drowned, but even at 350,000 BP, the Dover Straits may have been only one-quarter the width they are today.[17]

In the Wolstonian glaciation 160,000 years ago, the Scandinavian ice sheet dammed another lake in the southern North Sea. This time the water level did not reach above current sea level. Some natural sill in the basin was holding up the water and when it was overwhelmed an enormous flood carved the deeper Lobourg Channel through the Straits of Dover. Massive cascades and standing waves excavated deep scour channels along the 150-kilometre-long Hurd Deep (close to the north coast of Brittany), eroding more than 100 metres into the bedrock[18] (Fig. 57).

By the time of the last glaciation, rivers draining half the land area of western Europe were flowing through the Channel. Another ice-dammed lake in the southern North Sea formed around 20,000 years ago when sea level was 110 metres lower, from which the water eventually flowed out to the north, carving a channel 12 metres deep and 3 kilometres wide across Dogger Bank.[19]

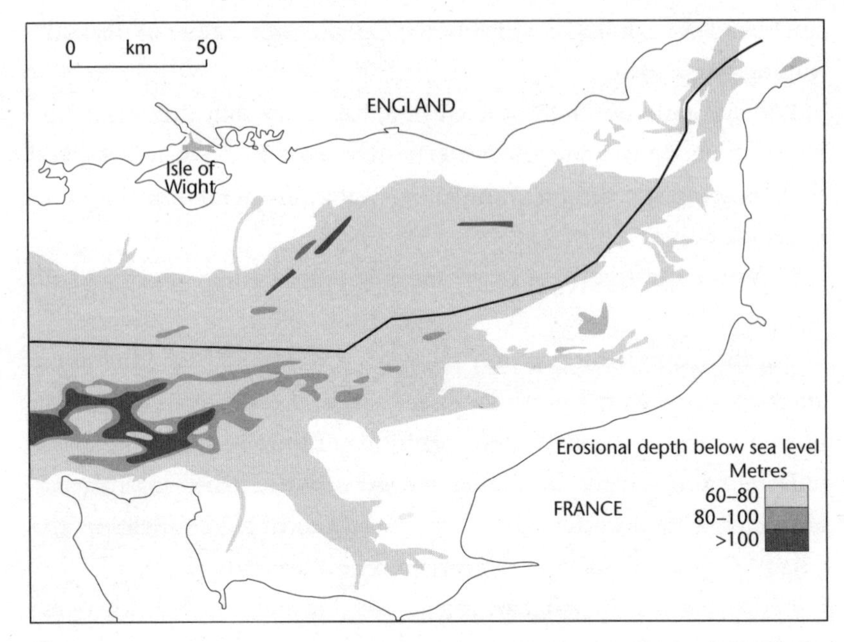

Figure 57. Deep erosion (in metres below current sea level) through the Channel after the second Ice Age south North Sea damburst

★

The one bright spot in the landscaping repertoire of the Ice Ages was their legacy of lakes and fjords.

Assisted by boulders embedded in the ice, a glacier gouges the base of the valley through which it flows. The thicker the glacier, the greater its power to dig. As a result, in approaching the melting 'snout' of the glacier, the base of the valley would get shallower. And then if the snout stayed in position, the melting ice would deposit mounds of stone and mud to form moraine hills. Take away the glacier and the combination of moraines with valley-floor gouging leaves a delightful legacy in lakes.

For much of the last glacial period, conditions of ice accumulation and melting stabilized around the English Lake District, with glaciers ending before the sea and therefore leaving lakes. The landscaping climatology was the same in Killarney and Snowdonia. Farther to the north in colder and higher Scotland, the glaciers carved deep valleys to lower elevations, which today form sea lochs. With even thicker ice sheets we get fjords.

We talk about glaciers 'retreating', as though, like a routed army, the ice scurried back into the mountains, but of course what really happens is that the ice starts to melt faster than it advances.

The ice overlay almost the whole of Ireland at 22,000 years BP and even by 18,000 years BP ice still covered over 60 per cent of the island. By 16,000 years the ice sheet had receded into the northern mountains, but Ireland was only ice-free by 13,000 years BP, by when sea-level rise had turned Ireland into an island. Before that[20] there was a tenuous natural causeway from Cornwall, with sill heights only a few metres above the local sea level, during which time elk, lynx, wolf, badger, and fox made it to Ireland.

*

A final phase of landscaping occurred in Scotland when, 12,900 years ago, a dramatic drop in temperatures led to a return of the Ice Age, and a new ice sheet accumulated across western Scotland in the 'Loch Lomond Readvance'. The conditions ameliorated a little until sudden warming 11,700 years ago. In Scotland the final stage of lake landscaping dates from this period. The moraines from the final ice-advance dammed a former sea loch to create Loch Lomond.

Further south around the Lake District more than sixty small mountain glaciers formed at this time, leaving their own legacy of mountain tarns but not disturbing the main radial lake architecture.

Along the coast of the south-west Scottish Highlands, for several thousand years during deglaciation, the land and sea were rising at much the same speed. During the arctic climate of the Loch Lomond Readvance, coastal ice floes and freezing abraded the rocky shoreline, carving caves, cliffs, arches, and sea stacks: a full student geomorphological textbook of features. The rock-cut platform is most conspicuous along the shores of the Firth of Lorn, where rising sea and rising land stayed in tandem for longest. Over the longer term the rising land has outpaced the sea. Inland of Oban the eroded shoreline platform is situated at 15 metres above today's sea level, tilting down to the west and south, reaching 11 metres above sea level north of Oban, falling to 4 metres in mid-Mull,[21] and intersecting modern sea level west of Mull at Fingal's Cave—itself originally eroded in this late glacial period (Fig. 58). To the south the shoreline is at 4.5 metres in north Arran and 1.5 metres to the south of the island. These contours reveal the shape of post-glacial rebound in western Scotland.

In the transit of people, plants, and animals, Britain stayed connected to the continent across the southern North Sea much longer than Ireland. Keeping on the north side of the Rhine, which now flowed out through the Dover Straits, it was possible to make the land journey across the southern North Sea until 6,000 years ago.

Rising sea levels presented an enormous challenge in a low-lying land. Water levels could advance rapidly, especially during northerly storms and king tides. We see today expanses of 6,000-year-old tree stumps, exposed where the sand or mud has shifted after a storm. When these were great living trees, the coast was everywhere backed by vast areas of dying salt-poisoned pine, alder, oak, and birch forests, while every estuary and bay was bristling with the standing trunks of dead trees. By 7,000 years ago the coastline had reached the Dogger Bank, which may already have become a large island, leaving nowhere to escape if the sea should suddenly advance. We know from sea-floor tools and encampments that this coastline was populated.

*

I had a project to investigate earthquake activity in Norway and was browsing a Swedish geology journal when I first encountered some grainy black-and-white photographs of fault scarps (surface exposures of recent fault displacement) from north Swedish Lapland. This was the last place one would except to see such images—far from a plate boundary, in the middle of ancient, stable crustal shield. Was this more of the Swedish exceptionalism that had outfoxed Charles Lyell?

I contacted the author of the paper, Robert Lagerbäck, an expert in Quaternary glacial geology at the Swedish Geological Survey, who

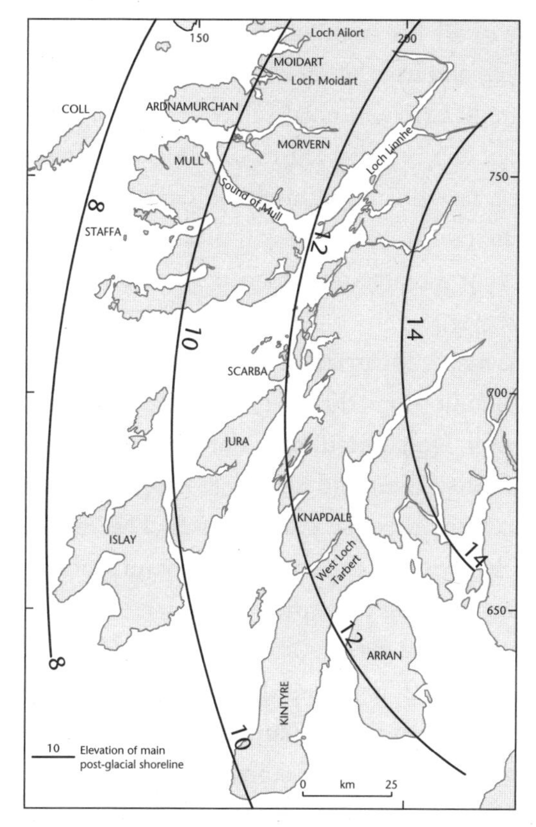

Figure 58. Uplift in metres of the raised wave-cut shoreline of western Scotland formed in the post-glacial period when sea-level rise kept pace with rebound

seemed to spend all his summers out in the wilds of Lapland. He suggested I come and have a look, so I arranged to fly up to Luleå, just south of the Arctic Circle in northern Sweden.

I checked out the boating jetties of the Baltic shoreline summer houses at Luleå, now marching uselessly over dry land. Some of the jetties ended in ramps, tilting down to a lower-level extension, to pursue the sea that was receding at a centimetre each year—the fastest in the whole region.

Robert took me in his grizzled Volvo a few hours' drive north to visit a couple of fault scarps that were accessible by dirt-track road and then a hiking trail. He pointed out a curious feature of the landscape: above a well-defined level, the hills were capped with 'hats' of much more vigorous trees than in the lowlands. This was the height to which the sea had eroded the soil out of the glacial moraines, just as the ice sheet was melting from the region.

Robert had arranged for the chartered helicopter he used for his remote field work to fly us along the longest and most impressive of these fault scarps, which he had named the Pärvie Fault—from a Lap word for a 'wave in the ground'. Above the tree line, the scarp of the fault could be traced for 160 kilometres south-west, riding along smooth ridges and cutting along the shores of lakes. It was the most amazing sight. We landed next to where the scarp was at its most impressive: a 10-metre-high overhanging rock wall (Fig. 59).

A few years later I returned to see the trenches that Robert had commissioned through the fault scarp of the more accessible Lansjarv Fault. He had also revealed sections through the abundant landslide deposits that had formed contemporarily with the fault scarps. The faults had broken around 9,000 years ago, just as the last remnants of the ice sheet were melting. In some of these landslides, Lagerbäck found the glacial till had undergone complete size gradation. The combination of liquefaction and violent vibration had size-sorted all the gravel and boulders. Since their first discovery, more of these fault scarps have been identified—one running for tens of kilometres through Norwegian Lapland, but none as long, or as spectacular, as the Pärvie Fault.

Figure 59a. The 9,000-year-old, end glacial Pärvie Fault scarps, northern Sweden, extends for 160 kilometres (above), exhibiting up to 10 metres displacement (opposite—with author for scale)

We know something extraordinary happened 9,000 years ago. The landslide evidence confirms these fault scarps were accompanied by intense shaking. The stresses that had been stored in the crust, over the tens of thousands of years the ice sheet remained in place, had erupted in a series of massive earthquakes as faults ripped through to the surface, raising great rock scarps cutting across the landscape.

Yet these faults did not look as if they had been caused by natural tectonic earthquakes, like the reverse fault scarp I saw in Algeria a few years earlier. They were too steep and too high.

The dimensions of the Pärvie Fault, its length and displacement, suggest an earthquake as large as magnitude 8.[22] Today in Sweden you might wait a hundred years to experience a magnitude 4 shock, one-hundred-thousandth the size. Researchers have looked for, but not found, equivalent fault scarps in the formerly glaciated northern United States and Canada. In Scotland there is some evidence, from landslides, of earthquake-shaking accompanying the end of the ice sheet, but no fault scarps like those of Swedish Lapland.

The shaking from these earthquakes would have been felt across the whole of north-west Europe. It seems possible one or another of

Figure 59b.

the later Lapland earthquakes was the trigger of the great Storrega Slide (dated to 8,200 years ago). A hundred kilometres offshore, the steep continental margin of mid-Norway, built out of glacial sediments, collapsed along a length of 290 kilometres. A 3,500-cubic-kilometre landslide of mud and sand slumped down the continental slope and flowed out into the deep ocean abyssal plain. This was among the largest landslides ever known. The slide triggered a tsunami that can be traced from contemporary sand deposits all around the regional coastline.

On the adjacent coast of Norway tsunami sands reach 11 metres above sea level. In the Shetlands the wave was a devastating 15–20 metres high. At coastal Montrose in north-east Scotland, the tsunami sand is a few metres higher than sea level. Arriving without warning, the tsunami would have overwhelmed shoreline communities all around the northern North Sea and flooded lowlands along the coast of the Netherlands and south-east England.

Beneath the centre of the Scandinavia ice sheet, the weight of the ice lowered the land by more than 300 metres. Take away the ice and the land slowly recovers. We talk about 'the dome of uplift', as measured by Charles Lyell on his visit to Sweden, but really this is a shallowing bowl that will eventually return to its original state.

Like a waterbed, the area around the ice sheet load rose up in compensation as some of the mantle rock beneath the crust was squeezed sideways. This rise of the land around the ice sheet is called the 'forebulge'. Just as the rebound of the land continues today, more than 10,000 years after the ice sheet melted, so the area of the former forebulge continues to subside.

The dome of uplift across northern Britain is not nearly as large, or rising so fast, as the dome which culminates in the Gulf of Bothnia, Sweden. Nowhere in western Scotland are the locals complaining about their sea disappearing or their harbours becoming too shallow, or demanding to relocate their fishing village. This is not because there is no uplift. The rise of the land at 1 or 2 millimetres each year is simply too little to be noticed over a lifetime. Nevertheless, a Viking harbour located at Oban, 1,200 years ago, would today be more than a metre shallower than when it was constructed.

Today the sinking forebulge continues to lower land levels across England, again at rates too slow to be noticed by port harbour masters.

While the rising rebound dome stretches the crust horizontally, the collapsing forebulge is in a state of compression. When that compression supplements the compression driven by plate tectonics, the forces can combine to trigger a fault to break.[23] This is almost certainly the explanation for big damaging earthquakes around the southern edge of the former North American ice sheet: in Charleston, South Carolina in 1886 and New Madrid, Missouri in 1811–1812, all located in the collapsing forebulge.

10

Up-land

In 1895 John Milne, father of earthquake science, founder of the first Seismological Society in the world (in Tokyo), and leader of the first government-funded mission for earthquake science, returned from Japan to search for a house. Rising nationalism allied with a fire that had destroyed his library convinced him to return to Britain after twenty years away, bringing his Japanese wife and his seismological assistant.

Milne's property search criteria were unusual. He had great affection for the living tectonics of Japan and had disparaged Britain as merely 'wearing away' while the land of Japan was renewed through tectonics. For Milne, 'location-location' meant proximity to earthquakes and he sought a house from which to run his seismic recorders, as close as possible to what he believed to be the most active fault in Britain.

Milne identified a house at Shide on the edge of Newport in the centre of the Isle of Wight, directly on top of the great fault and fold exposed to the west at Alum Bay. (His property search criteria could also have included a line of Devon villages from Torquay to Barnstaple, but in 1895 the significance of the Sticklepath Fault had yet to be appreciated.)

In an outhouse at Shide he installed one of his personally designed seismic recorders and sat back, hoping the local fault would deliver some tremors. His main quest at this period was to record the imperceptible vibrations of far-away earthquakes. Yet, over the next twenty years, the Isle of Wight Fault stayed shtum.

Was he unlucky, or had this most promising of faults gone extinct?

With something of the same romantic motivation, the great Victorian poet Alfred Lord Tennyson lived from 1853 until his death in 1892 in a grand house at Freshwater, the village where Robert Hooke was born, at the western end of the Isle of Wight. Every day Tennyson would aim to walk out west along the spectacular 'ridge of a noble down' (now named after him and the site of the Tennyson Memorial), rising up to 150 metres on the grass-covered vertical chalk promontory that leads on to the Needles, letting the sublime tectonic landscape feed his poetic imagination. 'One feels', John Betjeman wrote, 'that western Wight is an earthquake poised in mid-explosion and ready any day to burst its turfy covering of wild distorted downs.'[1]

Tennyson was the most scientifically literate of poets, a fellow of the Royal Society, avidly reading the latest works on geology and evolution. Living in a house on the seashore at Freshwater, listening to the grating roar of the shingle rolled by storm waves, he was intrigued by how sea levels and coastlines had shifted. T.H. Huxley, champion of Darwin's theories, honoured Tennyson as 'the only poet since the time of Lucretius, who has taken the trouble to understand the work and the tendency of the men of science'.

There rolls the deep where grew a tree
O earth, what changes hast thou seen!
There where the long-street roars, hath been
The stillness of the central sea.

Tennyson, 'In Memoriam', canto 123

Among all nineteenth-century scientists Darwin was the most self-deprecating and lucid writer. In his autobiography, the only exceptional talent he claimed was 'the power of observation'. Observation was indistinguishable from thinking. 'I cannot resist forming [hypotheses] on every subject.'

Charles Darwin brought his family to stay at Freshwater for the summer of 1868 in a house belonging to the renowned photographer Julia Margaret Cameron. We know Tennyson called on Darwin and Darwin on Tennyson.

While in the vicinity, the Darwin party made two day-trips to the vertical strata at Alum Bay (commended as 'very grand')[2] and another to visit the Needles. However, his days as a geologist were over.

Tennyson, Darwin, and Milne had all been attracted by the evidence of 'recent' tectonic activity, but the Isle of Wight was no longer the tectonic front line.

★

The ocean boundary faults and rifted basins to the south and west of Ireland are dormant. All that is exposed of the Rockall Bank, a submerged micro-continent, is an absurd barn-sized rock. This last remnant of a Paleocene volcano and bulwark against the force of Atlantic waves is made of the toughest microgranite, just like the rock of Ailsa Craig. There seems little prospect Rockall would be quarried for curling stones.

Rockall Bank has no source of sediment, so the ancient surface, dry land 40 million years ago, has sunk lower than the lowest Ice Age sea levels. In the direct path of the colossal tsunami that began the Cenozoic, as the land recovered through the Paleocene there must have been forests, birds, beaches, outcrops, a whole living landscape which, over millions of years, all slowly passed beneath the waves. Much more recently huge icebergs became grounded on the drowned plateau, scarring the seabed.

To the north-west of Scotland the big reverse faults of the Miocene also appear parked. The spreading in the Norwegian Sea eventually became unblocked and renewed, and no longer drives reverse faults along the continental margin of Scotland and the Faroes (Fig. 46).[3]

⋆

Surprisingly, the oldest element in the geography of north-west Europe is the North Sea. A shark that swam into the basin today would find a sea not unfamiliar to its Cretaceous ancestors 100 million years ago. Nothing else of this scale in the region is quite so old.

Two hundred and fifty million years ago a rifted depression ran from northern Germany to the coast of Yorkshire. In the middle of the Pangaea supercontinent, evaporation in the hot desert climate produced 500 metres of salt.

Rifting in the North Sea became renewed as Pangaea split apart from 200 to 120 million years ago, and the Atlantic Ocean began to form. Phases of rifting and extension were followed by broader subsidence.

Through the Cenozoic the North Sea basin has continued to sink, filling with 480,000 cubic kilometres of sediments eroded from the surrounding lands (equal to 2 kilometres of erosion from the whole British land area).

Most extraordinary is what happened at the end of the period. Through the Pleistocene, the North Sea floor started to sink much faster than before. The basin developed a central 600-kilometre-long NNW–SSE-trending depression now more than 1,250 metres deep, a trench which, as it sank, has filled with sediments eroded by the ice sheets (Fig. 60). Over much of the North Sea, more sinking has happened in the past 2.5 million years than in the whole of the previous 65 million years of the Cenozoic. The maximum thickness of Paleocene to Pliocene sediments is 3,000 metres. The basin has lately been sinking

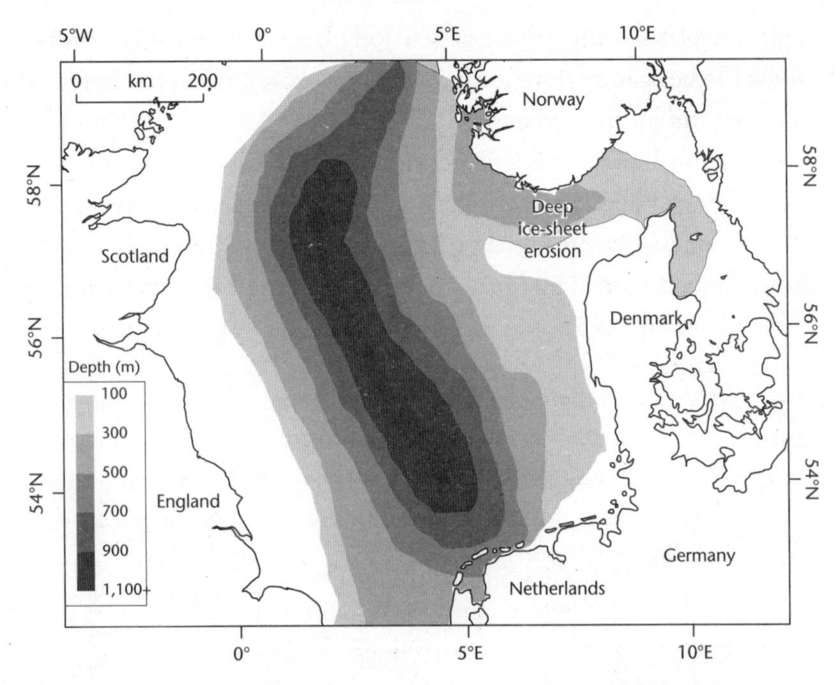

Figure 60. Depth to the base of the Pleistocene sediments in the North Sea

at 500 metres in a million years, ten times faster than the average for the whole Cenozoic.

The Dutch have recklessly settled where the River Rhine delta has expanded over the subsiding southern North Sea. Digging a well in search of a firm foundation in Amsterdam is a bottomless task. Buildings in Amsterdam have to be deep-piled into the thickest underlying sand layer. There is no bedrock.

However, pass into southern Holland, and the subsidence transforms into active rifting on NW–SE faults in the Lower Rhine Graben. The crust has widened more than 100 metres in the Pleistocene.[4] This tectonics is still active: the largest twentieth-century earthquake broke beneath Roermond, Netherlands, in 1995 with a magnitude of 5.4. In 1756 an earthquake close to the city of Duren had a magnitude of 5.7. Larger earthquakes, with surface displacement, have occurred on these faults, although apparently not in the last 10,000 years.[5]

In the plate tectonic toolbox we don't have an easy explanation for what has been happening. But there is much more that has happened around the North Atlantic that does not have a plate tectonic explanation.[6] In Norway there are two elongated domes of uplift, hundreds of kilometres across: one centred on the 2,400-metre-high Jotunheimen ('home of the giants') mountains of central Norway and a second with a culmination in coastal northern Norway, both of which have largely risen over the past 30 million years (Fig. 61).

I first travelled to Norway on a car ferry from Newcastle upon Tyne, and was struck by the contrast in coastlines. Leaving the

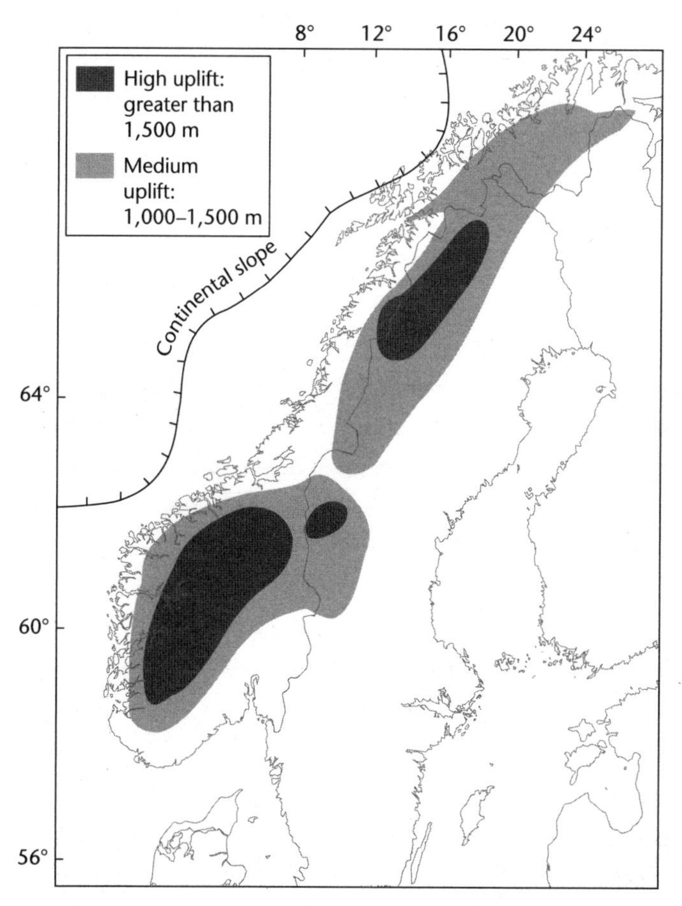

Figure 61. The 'quiet' Neogene uplift of Norway

industrial lowlands in the evening, you woke up to naked, ice-worn, gneissic, 'skaergaard' islands and precipitous mountains at the back of Bergen. Seismic reflection profiles show that the eastern side of the northern North Sea has been raised up, carrying with it the faults and sedimentary basins. The topography of westernmost Norway comprises the same metamorphic basement rocks that underlie the thick piles of sediment in which Norway has found all its oil and gas wealth offshore. The mountains of the Lofoten islands are so steep because these were once the base of rifting faults.

All the uplifted Mesozoic sediments that once overlay this basement have been pillaged by the ice. Yet, not quite 'all'. While working in northern Norway I took the opportunity to make a unique secular pilgrimage to the far north island of Andøya where there is the only surviving Cretaceous/Jurassic sediment on-land in Norway,[7] just above the tideline, a nondescript grey-brown sandstone, from which I took a memento cobble (long lost in one house-move or another).

Before the Cenozoic, east Greenland was low-lying. Today at 67 degrees latitude the general height of the basalt plateau is more than 2,500 metres, and the top of the dome reaches more than 3,500 metres.[8] Somehow, beneath both Norway and Greenland, massive amounts of lower-density mantle-derived material have arrived to create a cushion that has raised and domed the overlying continental crust. In Norway this has happened without any accompanying volcanic activity or rifting. Norway and Greenland have emerged while the Rockall Bank has submerged.

★

Two hundred and fifty years ago, Gilbert White of Selborne, Hampshire, was taken by the poetic power of the sunken lane: 'Among the singularities of this place, the two rocky hollow lanes, the one to Alton and the other to the forest, deserve our attention', worn down 'by the traffick of ages and the fretting of water'.[9]

We know, from our own steps, how small a trace we leave: a kicked stone here, a rut of a mountain-bike there, from which we can infer the millions of people and animals, the 'traffick of ages', who inadvertently carved this artisanal ravine, over hundreds or even thousands of years.

Many sunken lanes have been tarmacked, neutering their ability to grow any deeper, but some of the finest holloways in Sussex, Dorset, and Gloucestershire are still alive, acquiring stream-beds, actively eroding through the deep layers of clay and sand: dark, overgrown tunnels, 10 metres or more deep.

The sunken lanes take me back to Tolkien's Shires and the hobbit houses, cut into the soft earth of the valley walls. If a ravine can be eroded by footfall, if the badgers can dig their setts deep, we are in the softest geology of clays and sands.

How is it that such a soft and vulnerable stratigraphy is found in the Dorset hills 200 metres above sea level? To have endured, these fragile landscapes must be very young.

Once again we are looking for geological clues.

Fortunately we have a distinctive marker horizon: the marine 'Red Crag'—the shelly sand formation that underlies eastern Norfolk and Suffolk. Red Crag was deposited 2.5 million years ago (around the start of the Pleistocene), in a shallow sea, perhaps 20 metres deep, out of reach of storm waves. Inland in Suffolk, the Red Crag sediments infilled deeply scoured NE-trending 'tidal' channels. Heavy minerals mixed in with the sands show the sediments were sourced from the great 'Eridanos' river delta that, before the Ice Ages, emerged from the Baltic, spreading across the North Sea into East Anglia.[10] There was little sedimentation from English rivers, probably because the contemporary coastline was far to the west and the land remained thickly forested.

Close to Norwich, for every kilometre one passes to the east the base of the Red Crag descends by 2.5 metres. Offshore that steepens to 3.5 metres for each kilometre.[11] Inland across East Anglia the base of the Crag deposits is tilted up to the south-west at a gradient of around 1 metre for each kilometre, even as the surviving outcrops become more fragmented (Fig. 62). The base of the Red Crag reaches

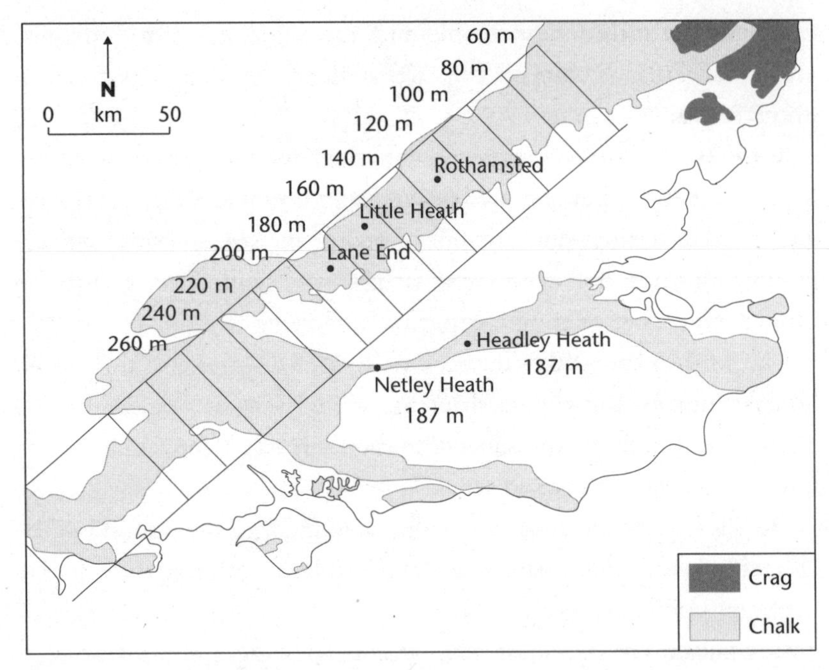

Figure 62. The tilted uplift of the Red Crag north of London, and the more consistent uplift of the Weald, since 2.5 Ma

50 metres above sea level at Stowlangtoft, near Bury St Edmunds, and 90 metres at Stansted Mountfichet close to Saffron Walden. Farther to the west at Rothamsted in Hertfordshire, fossiliferous Red Crag sands are found at 130 metres above sea level[12] rising to 170 metres at Little Heath.[13] Still rising, 28 kilometres to the south-west of Little Heath, we find Red Crag sands and pebble beds (derived from the marine erosion of the Paleocene Reading Beds) at 190 metres above sea level, exposed in trenches dug for water mains in a wood next to the M40 at Lane End.[14] And that is the most westerly outcrop known of the Red Crag.

Yet, given that the tilt continues, if we project the base of the Red Crag farther to the west we find ourselves flying over the landscape of southern England, grazing White Horse Hill at 261 metres, the highest point of the Berkshire Downs, and almost clearing 277 metres at Liddington Castle near Swindon.

However, along the Chilterns scarp to the north, the hills rise at least 70 metres above this Red Crag surface, to a high point at 267 metres on Haddington Hill near Wendover. There must have formerly been chalk cliffs, maybe even a chalk island, fringing the northern side of the Pliocene sea.

Pockets of Pliocene marine sediments (the 'Lenham Beds') are exposed at 187 metres above sea level in north Kent: sands rich in iron preserved in fissures in the chalk, indicating this was once a sea floor. At Netley Heath in Surrey the iron-rich sands contain Early Pleistocene 'Red Crag' age fossils. The Pliocene sea seems to have covered the whole of the eroded Weald anticline.

Over northern France isolated outcrops of Pliocene marine sediments are found at 150 metres elevation, south of Dunkirk[15] and on the incised plateau of the Lower Seine.[16]

All these outcrops help map the shape of uplift since the Pliocene. The tilted surface that runs inland from East Anglia becomes a more uniform uplift south of London which extends into Normandy. The tilting must eventually crest, beyond which the land must dip down to the west. Where is the culmination? Over Dorset? Over Bristol and Somerset? Does the crest of uplift reach 250 or maybe 300 metres? Did the 400-metre-high Exmoor Plateau once continue uninterrupted across to south Wales? Farther west, the westerly tilt supplemented the slope of the Cornubian granite cupolas, from 600 metres high in Dartmoor down to sea level in the Scilly Isles.

The Cotswold hill plateau tilts up to the north-west, where the Jurassic limestone hills are incised by deep valleys, linked to the dramatic reduction in elevation of the River Severn riverbed through the lower Ice Age sea levels. The tilt has been explained as isostatic rebound, a result of the unloading. The Cotswold plateau formerly extended far to the north and west. Where the plateau was deeply eroded and unloaded, the crust has risen, 'rebounded', by between 40 and 60 metres.[17] The highest summits follow the edge of the escarpment, with views across the Severn Vale, where the Cotswolds formerly continued, towards the Forest of Dean and into Wales. Once the limestone cap was pierced, the underlying clays were washed by the rain and subject to spectacular landslips. In the valleys around Stroud, steep streams carved into the plateau powered scores of watermills.

One-sixth the scale, and smothered in fields, trees, and houses, this is 'England's Canyonlands'.

Yet the most sensational part of this story is that, against all expectations, only 2.5 million years ago almost all of southern England lay underwater. Global sea level at the start of the Pleistocene was no more than 20 metres above current levels and with the shelly sands deposited in 20 metres water depth the current height of the Red Crag formation reveals approximately by how much the land has risen.

The landscapes of Dorset, Kent, and the Cotswolds are as newly sculpted as the volcanic ridges of the Coromandel Peninsula, New Zealand.

Denmark and southern Scandinavia have also emerged—both in the Miocene and more clearly in the Pliocene.[18] After the land had risen several hundred metres, even as much as 1,000 metres in northern Jutland, the sea eroded back the uplifted land.

As though on a see-saw, over 2.5 million years while the North Sea floor has been sinking England has risen, by about one-quarter of the amount by which the floor of the North Sea has sunk. (Inland of East Anglia the rise has been 1 metre per kilometre, offshore the fall 3.5 metres per kilometre.) This is an idea that would have resonated with Darwin. Having witnessed uplift on the coast of Chile, he believed there must be compensatory sinking of the Pacific ocean floor (with tropical islands sustained at sea level by sunlight-chasing corals).

Some part of this 'England–North Sea see-saw' has come from isostasy. Erosion of the raised landscape has reduced the load on the mantle and, like a melting iceberg, the land has partly risen in compensation. The sediment load that has accumulated in the subsiding North Sea has weighed down on the mantle and increased the sinking. Yet isostasy has only amplified the tectonics.

*

Is England still rising? How would we know?

One test comes from finding the elevation of beaches formed by higher sea levels, during warm interludes between the glacial periods.

Nine hundred thousand years ago, at Happisburgh, on the north-east coast of Norfolk, five humans, including children, left fifty footprints in the sand. They took less than a minute to walk past before a turn of the tide, or river flood, covered the impressions with another sand layer. Their path ran alongside a distributary channel of the proto-River Thames. After 900,000 years in darkness the footprints re-emerged within an eroding cliff. Exposed for a few days in 2014, and then they were gone. At this location, sea level looks to have been similar to today.

This fleeting episode marks the beginning of an on-off relationship of arrivals and departures that continues to the present day. This group, or their ancestors, arrived in Happisburgh by migrating around the coast of the southern North Sea. They were not to know it, but they were on a peninsula of north-west Europe.

Britain first became an island 450,000 years ago, when the great southern North Sea lake overtopped the hills that ran from Kent to the Boulonnais. Since that time, Britain has oscillated in and out of insularity. At times when the climate across Britain was arctic, sea levels were tens of metres lower and Britain was once again a peninsula of Europe. Animals walked across what is now the southern North Sea, on dry land between one great river flowing north and another flowing west through the Dover Straits. The animals that made this journey were suited to colder conditions: lemmings, mammoth, horses, woolly rhinoceros, and reindeer. Meanwhile, early humans removed to live in caves decorated with animal art in the less hostile climate of southern France.

As the climate warmed and sea level rose, Britain was once again an island.

Over the past half a million years there have been three episodes when sea levels were higher than today.

The highest sea level of all was around 404,000 years ago (known globally as Stage 11,[19] or 'Hoxnian' in the UK[20]). Along coastlines we believe to have been stable, like South Africa and the Bahamas, Stage 11 seas reached 13 metres above current levels.[21] (To raise global sea level by more than 12 metres requires melting both the Greenland and West Antarctic ice sheets.)

In 1924, while digging a trench for a sewer at Bembridge School above cliffs at the eastern tip of the Isle of Wight, a layer of clay was exposed with seams of peat containing the pollen of spruce trees and arctic buttercup. The peat had formed in an estuary during the Hoxnian interglacial. The most surprising feature was the elevation of the deposit: 38–40 metres above sea level.[22]

Another raised beach formed in the same interglacial was traced from Slindon, Sussex for 40 kilometres west into Hampshire, backed by a former cliff-line, now largely obscured by landslides. Although known as the '100 foot beach', it is actually 40 metres above sea level, the same elevation as at Bembridge.

Fossils from this raised beach exposed in a gravel pit at Boxgrove reflect the temperate climate of an interglacial. Overlying the sands is a clay containing rodent teeth and flint hand axes.[23] Humans had evidently lived on the beach soon after the time of the highest sea levels. In 1982 an excavation[24] uncovered animal bones including lion, bear, rhino, and giant deer, many of them carefully butchered for their meat. In 1993 a human (*Homo heidelbergensis*) tibia bone was found, at the time the oldest human remains in Europe. The climate worsened, sea level fell, and this is the last time we have evidence of the human occupation of a British beach, until the end of the latest glaciation.[25]

Since that time the land around the Solent has risen by at least 27 metres. Or to put it another way: these early shoreline humans entered the elevator of prehistory at ground level and left it (400,000 years later) on the ninth floor.

Before the most recent glacial period, the latest interglacial is known in Britain as the Ipswichian, and internationally as the Eemian or 'Stage 5'.[26]

No humans returned to island Britain in time to beachcomb next to the sea at the high water mark of the Eemian interglacial. However, many other species of animals did make the journey: the wood mouse, the straight-trunked elephant, the hippopotamus, and the fallow deer.

They enjoyed a reasonable climate, empty beaches, plentiful seafood, and no human predators. By the time of the warmest interglacial climate, the Channel had become a formidable moat.

How high was the global Eemian sea level? A meticulous study of stalactites in eight coastal caves on 'stable' Mallorca shows that sea level rose to 6 metres above current levels at 127,000 years ago, gradually falling to 2 metres from 122,000 to 116,000 years ago BP.[27]

The height of Eemian age beach deposits can therefore reveal land-level changes over the past 127,000 years, although we may be more likely to find raised beach evidence from the longer-lasting +2 metres elevation sea level than the short-lived +6 metres.

In the Netherlands, continued North Sea sinking has lowered Eemian age sediments to 10–20 metres below current sea level.[28]

At Stutton, Suffolk, on the northern bank of the Stour estuary, the highest recorded Eemian sea level is today 1.2 metres underwater.[29] However, around the coast of southern England there are many Eemian raised beaches, banks of sand and gravel, or cave deposits, now out of reach of the sea. At Black Rock, Brighton, the raised beach shingle rises to 12 metres above sea level. In the vicinity of Selsey and Chichester there is a 7.5-metre-raised beach. At the Berry Head limestone headland in Torbay there are raised beaches in caves: one identified as Eemian at 5.8 metres. There is a '5-metre beach' located in bays around the Channel Islands whereas Belle Hougue cave on Jersey has an Eemian raised beach at 8 metres.[30]

The raised rock-cut platform, assumed to be from an Eemian sea level, is exceptionally well preserved at Prawle Point, the southernmost tip of Devon, looking like a product of far more recent tectonics. We should take particular note of the Eemian; it could, given global warming and ice-sheet melting, be the sea level we are bequeathing our great-great-grandchildren, alive in the twenty-second century.

Between the Eemian and the Hoxnian there was another prolonged interglacial known as 'Stage 7' with two episodes of high water levels, one around 200,000 and a second 240,000 years ago.[31] The actual sea levels reached are not so well agreed, but probably at least 3 metres above current levels.[32] At Portland, Dorset, the Stage 7 raised beach deposits reach up to 16.2 metres.[33] At Bembridge, on the Isle of Wight, another Stage 7 raised beach extends up to 18 metres above sea level.

Broadly all the interglacial raised beach deposits of southern England reveal uplift since they were laid down, although that may be less certain for the Eemian levels.

Only one raised interglacial beach deposit has been identified from northern England.[34] Cross the railway on a path at Easington, County Durham, and scramble down a gully to a ledge in the cliff, where 33 metres above modern sea level there is a fossilized beach deposit. The Easington raised beach has survived because the cliff face is of tough magnesian limestone and is sheltered from storm waves. Age dates cluster from 150,000 to 250,000 years BP, indicating a Stage 7 interglacial age.

The Durham coast has risen by up to 30 metres over the past 200,000 years. That is almost twice as fast as West Sussex and the Isle of Wight.

★

The third line of evidence for 'Up-land' comes from 'river terraces', on the flanks of the largest valleys. Terraces are carved by rivers in flood, eroding the edge of the valley, and depositing gravel and sand. If the land is rising and the river is eroding, the older terraces are abandoned and stranded.

The broadest terraces formed during glacial periods, when rivers were loaded with debris flushed by the spring thaw. In the tundra climate there was little vegetation to bind the soil.

Dating the gravels can reveal how far the river has cut down since the terraces were formed.

Each section of terrace, each gravel, gets some local name: the 'Nettlebed Terrace', the 'Boyne Hill Terrace'. Along the broad valley sides, there can be dozens of terrace deposits, each formed in some short episode in the history of that river's past. The challenge is to link terraces of the same age along the length of the river.

The terrace gravels high along the valley sides of the River Thames preserve a detailed record for more than a million years and include formations like the Kesgrave Gravels, deposited before the 450,000-year-old Anglian glaciation[35] diverted the river to the south.

From the age and height of the terraces we can determine how far, and how fast, the river has cut down since that terrace was formed. The terraces reveal that the Thames has eroded down[36] by 200–250 metres since the beginning of the Pleistocene (2.5 million years ago), with the uplift increasing westward. Of this total, for both the Thames and the Solent rivers, a consistent 70 metres has occurred over the past 870,000 years,[37] rising to 80 metres along the River Frome at the western end of the Hampshire Basin.[38]

All the evidence of former water levels revealed by raised beaches, abandoned river terraces, and the Red Crag uplift surface tells a remarkably consistent story. For the past 2 million years, central southern England has risen 70–100 metres in a million years, continuing across the Channel to 55–60 metres uplift in the last million years around the Lower River Seine.[39]

In the soft geology of clays and sands across south-east England, land uplift leads to broad valleys and stranded river terraces. In passing to the west and north, where the bedrock is older and harder, the signature landscape of uplift is the 'incised meander' (Fig. 63).

The incised meander requires two ingredients. First, it takes a low-gradient river flowing across a flat landscape, typically close to sea level, shifting its meandering path in the flood plain. ('Meander' is a

Figure 63. Incised meander on the River Wye

river in western Turkey, curling aimlessly through a broad valley that is sinking from tectonics and choked with sediment.)

Second, the land rises (or the sea level falls) and the river cuts down. Where the bedrock is resistant to erosion the latest meanders become incised, as though chiselled by a stonemason. The ephemeral process is made permanent.

Incised meanders are found in an arc running from the Lower Tamar Valley in east Cornwall, through north Devon (the Torridge and Taw rivers) and the Lower Dart in south Devon, on to the famous curves of the 200-metre-deep Lower Wye Valley between Ross and Chepstow. Incised meanders reappear in north-east England, along the River Wear in Durham and the River Wharfe. The ice sheets likely destroyed many other examples across western and northern Britain. All these rivers were close to sea level in the Pliocene before the land began to rise.

Across the Channel, another series of incised meanders runs from north-east France into the Ardennes. Most famous are those along the Lower River Seine, starting in Paris. Cliffs along the river expose pinnacles of chalk and can be more than 100 metres high.[40] The Semois River south of the Ardennes in Belgium follows a serpentine gorge, as does the Mosel along the border between France and Germany.

In the late eighteenth century, incised meanders were thought to be the epitome of the 'picturesque'. Bolton Abbey on the incised River Wharfe was the subject of a painting by J.M.W. Turner (who imagined precipitous cliffs on the valley sides). The river bends around, and the walls of the valley enclose a sheltered space, with a balance of woodland, river, and meadow.

For Wordsworth the incised meanders of the River Wye inspired his 'Lines written a few miles above Tintern Abbey': an existential riff on connectedness and spirituality:[41]

> ... And again I hear
> These waters, rolling from their mountain springs
> With a soft inland murmur. Once again
> Do I behold these steep and lofty cliffs,

> Which on a wild secluded scene impress
> Thoughts of more deep seclusion, and connect
> The landscape with the quiet of the sky.

Fifteen hundred years earlier, the sentimental Roman poet Ausonius wrote a lengthy paean to the twisting Mosel in its deep incised valley, as the finest of all rivers ('all other rivers will bow to the Mosel'[42]). The incised River Rhine is today a UNESCO World Heritage site.

At Durham, more practically, the tight curves of an incised meander have created the natural fortifications of a well-defended citadel of cathedral and castle—an identical configuration to the Swiss city of Bern.

Surveyed over hundreds of thousands of years we see consistent evidence of uplift across England. After the local ice sheets, rebound and dramatic changes in sea level, does this uplift continue today?

★

The eighteenth-century rock-cut water-height markers in the Gulf of Bothnia (see Chapter 9) were only usable because there was no tide. Where the twice-daily tides go up and down several metres, shifting through neaps and springs, there was no simple way to record a water-level datum. How about the rare extremes?

In central London, in response to each catastrophic storm-tide flood, the walls that line the river were raised. Close to Blackfriars we have four ages of walls, one on top of another: in 1879 (following the 1874 flood), 1900 (following the 1897 Thames floods of east London), 1928 (following the extensive January 1928 flooding of Westminster), and 1955 (following the 1953 east coast flood). Overall the walls raised the protection by more than a metre.

For the hundred years to 1960, tide gauge measurements reveal that extreme high water levels in central London had risen 82 centimetres, far faster than regional sea levels. That seemed to leave only two explanations: either the land is sinking or storms are growing stronger.

The sinking explanation seemed puzzling. In contrast to true delta cities, like Venice, Amsterdam, or Bangkok, London is founded on well-compacted 54-million-year-old London Clay.

Were storms growing stronger? On the Dutch coast the record high water level occurred in a storm in 1570. Unlike Atlantic hurricanes, there is no evidence for a modern trend towards stronger storms.

The reason for raising the flood walls turns out to be neither subsidence nor storminess. Instead it was all a consequence of dredging.

For the Romans, London represented the nearest point to the sea where the Thames could be forded, inland of the tidal estuary. The Romans built a wooden bridge across the shallow river, requiring boats to unload (and thereby creating the reason for the city). When the first stone 'London' bridge, with its nineteen arches, was constructed and opened in 1209, the downstream tidal range may have been 1 or 2 metres. The narrow arches of the bridge acted like a weir, so the tides scarcely advanced upstream.

Old London Bridge was finally demolished in 1831. A teenage Charles Dickens used to come and sit on the bridge to watch life on the river, from which it became a young David Copperfield's 'favourite lounging place'. You can still see the rapids over the rubble on a strong ebb tide.

As the new London Bridge with its broad spans was opened 40 metres upstream, there were a number of unforeseen consequences. The tidal range (4.6 metres in 1799) expanded both upstream and downstream. Below London Bridge boats found themselves stranded at low tide.[43] Upstream low tide level dropped by a metre and the faster tidal currents eroded the riverbed by 1.5 metres, undermining the foundations of the next bridges upstream at Blackfriars and Westminster. Even after the opening of the new London Bridge, as the riverbed eroded, the tidal range expanded from 5.1 metres in 1832 to 6 metres by 1834 (and 6.3 metres in 1877).

Low tide levels at Teddington, 25 kilometres upstream, lowered by 80 centimetres. A boat could now float from Putney to Richmond in a single rising tide rather than having to be laboriously towed by a horse. Along Chiswick Mall, and in Richmond, fancy Georgian river-fronting houses discovered the neighbouring Thames had become tidal, flooding their parlours in spring tides.

The loss of the old bridge meant the end of frost fairs. Every twenty or thirty years through the colder Little Ice Age years of the seventeenth and eighteenth centuries, the river at Westminster froze. Old London Bridge dammed the freshwater upstream and kept the level stable so that thick ice could form. The last frost fair was in January 1814. After 1831 the water at Westminster was always brackish, fast-moving, and changed level by several metres through the tides.

Another lost tradition: the opportunity after a summer drought to wade across the knee-deep, sewage-ridden Thames, upstream of London Bridge.[44]

Through the nineteenth century, London tried to maintain its role as the principal port of the region. Huge docks were constructed to cope with larger ships. The estuary had to be dredged, deeper and deeper. Between 1909 and 1928, dredging lowered the downstream estuary riverbed by 2 metres. As a result high tides in central London increased by 35 centimetres.

Today, even in the absence of a storm, the maximum tidal range at London Bridge is 8.01 metres, 2 metres more than in Dickensian 1834. Increased 'range' means lower neap tides and higher spring tides. The water level reached by the same combination of wind-driven storm-surge plus high tide would today be a metre higher than in the 1830s.

The raised London flood walls tell us nothing about tectonics. They reveal the desperate attempt to hold onto London's dominance as a port. The race was finally lost in the 1950s, as large container ships diverted to Tilbury, Felixstowe, or Rotterdam. After the 1960s, if all the dredged sediment had been returned to infill a rewilded Thames estuary, restoring its medieval marshiness, central London storm-surge flooding would have gone away. Instead, noting the 80-centimetre increase in flood levels over the previous century, we got the Thames Flood Barrier, started in 1978 and built at a cost of £478 million (around £3.5 billion in 2024 money).

In fact the earliest record we have of a British sea level comes from the most catastrophic British coastal flood of the past 500 years, on 30 January 1607. Flooding extended across the low-lying land on both coastlines of the Bristol Channel from Barnstaple to Gloucester to Cardiff and took an estimated 2,000 lives. The catastrophe was caused by a wind-driven storm-surge superimposed on an astronomical tide that oceanographers calculate was among the highest of the seventeenth century. (It was not, as widely and misleadingly claimed by the BBC among others, a tsunami!)

In Barnstaple contemporary reports identify the specific streets flooded, from which we calculate the water reached 6.15 metres above sea level (Ordnance Datum). However, the locals reported the flood was '5 to 6 feet' (1.5–1.8 metres) higher than any previous tide level (which had to be at least the highest astronomical tide). Modern tide tables calculate the expected astronomical high tide at Barnstaple for 30 January 1607 as 5.7 metres. Add on the '5 to 6 feet' and we reach 7.2–7.5 metres: more than 120 centimetres higher than observed.

The maximum height of the water was also memorialized in several flooded churches. Not far from Avonmouth, the flood height was recorded with a brass plaque at Kingston Seymour church (surveyed to be 7.74 metres above sea level). At nearby Avonmouth the predicted astronomical tide on that morning is calculated to have been 7.86 metres. As with the situation at Barnstaple, to have so overwhelmed flood defences and swept away so many houses, livestock, and people, the actual water levels reached that day must have been at least 1 or 2 metres higher than expected.[45] In Bristol the morning tide that day was said to have been '9 feet' higher than the evening tide. We have a paradox: the levels reached in the most catastrophic sea flood in Britain's history seem no more than a modern spring tide.

Both at Barnstaple and close to Avonmouth it seems that extreme sea level has risen by at least a metre over the past 400 years. Some part of this is sea-level rise. Another component may be an expanded tidal range, but it also appears that the land has been sinking. At Newlyn in Cornwall the measured sea level has risen 1.7 millimetres per year since 1916,[46] extrapolated to more than 70 centimetres since 1607. Less than 1,000 years ago, before at least a metre of local sea-level rise, all the Isles of Scilly were conjoined into a single island.[47]

How much of 'sinking England' is caused by the collapsing forebulge from the ice sheets? Continuous GPS station measurements indicate southern England is lowering at 2–3 millimetres each year. This contrasts with uplift of between 1.3 and 1.5 millimetres per year in central Scotland.[48] England is in the sinking forebulge from both Scandinavia and Scotland. Even the Outer Hebrides are in Scotland's sinking forebulge. On North and South Uist, Neolithic chambered cairns are flooded by high tides.[49]

11

Quake Britain

So far in this story we have considered the principal geological faults that weave their way through Britain—and by what distance their sides are 'offset'. It requires a leap of imagination to conjure up the thousands of shattering earthquakes that accompanied the movement along these faults.

To switch between these two worlds—the geological grammar of faults and the seismological chatter of earthquakes—requires translation. We need a 'Rosetta Stone' to guide the conversion.

In 1935, in order to communicate to local Los Angeles journalists the size of the latest tremor, the seismologist Charles Richter devised the earthquake magnitude scale. Richter's magnitude was calculated from the amplitude of the earthquake vibrations measured on a standard seismic recorder, adjusted for the distance from the source.

As there was such enormous variation in the size of these records, from the tiniest squiggle to wild oscillations that sent the pen leaping out of the recorder, he took the logarithm of the amplitude as his measure. One step on the Richter magnitude scale, it turns out, makes a thirty-fold difference in true size: a magnitude 6 earthquake releases thirty times the energy of a magnitude 5.

By the 1960s Richter's scale was discovered not to work for the largest earthquakes, above magnitude 7. In response Japanese seismologists found a better measure of earthquake energy release (termed 'seismic moment'), but, recognizing the currency of Richter's scale, they converted seismic moment back to match Richter's original scale as 'moment magnitude'.

For consistency we will try to use moment magnitude. Meanwhile, national seismological agencies have continued to report a local magnitude (Ml), which tends to be in the range of one-third to half a unit higher than the moment magnitude.

For many smaller events before 1970, and even larger shocks before 1900, there are no useful seismic recordings. Instead, we have to estimate the magnitude from the area that experienced a given level of shaking, calibrated against recent earthquakes from the same region.

Seismic moment is calculated from multiplying together three ingredients: the dimensions of the fault that breaks, the amount of displacement on that fault, and the strength of the rock. Moment magnitude is abbreviated to 'Mw'. As rock strength is fairly consistent through the crust, the ratio of the fault displacement in a single earthquake to the length of the fault rupture stays in a narrow range: around one part in 10–20,000. This allows us to outline an approximate relationship between earthquake size and the area of the fault that breaks.

These are some approximate conversions:

A magnitude 3 earthquake requires an area of 0.1 square kilometres (10 hectares) of fault to break,

A magnitude 4 earthquake requires 1 square kilometre,

Magnitude 5: 10 square kilometres,

Magnitude 6: 100 square kilometres,

Magnitude 7: 1,000 square kilometres,

Magnitude 8: 10,000 square kilometres, and

Magnitude 9: 100,000 square kilometres.

To give these areas some context, the area of Hyde Park and the adjacent Kensington Gardens in London is 2.5 square kilometres, equal to the fault rupture of a magnitude 4.2 earthquake.

The area of the Isle of Wight is 380 square kilometres—equal to the fault that can generate a magnitude 6.2 earthquake.

The area of Wales is 21,000 square kilometres—equivalent to the rupture area from a magnitude 8.1 earthquake.

For typical fault ruptures, the Hyde Park-sized fault shifts 5 centimetres, the Isle of Wight-sized fault by 1 metre, whereas the Wales-sized fault shifts by 10 metres.

Even the pitch of the vibrations from the earthquake reveals the size. The larger the fault, the lower the frequency, like the longer string on the double bass as compared with the violin. The earthquake is the roar as the crust fractures and slips. Short fault ruptures scream, longer faults growl as they break. The longer the fault, the louder and deeper the vibrations.

*

We have reports of British earthquakes going back more than 1,000 years. In medieval times, apart from thunder, there was nothing else with which the vibrations could be confused: no sonic booms, rumbling lorries, or quarry explosions. The shaking from an earthquake was enough to wake a monk from his contemplations. In case it might augur some miracle, he should list the experience in the monastic annals.

Yet before 1600 the surviving information is generally too fragmentary to map the earthquake's location and size.[1]

For a century of Plantagenet kings (through the twelfth and thirteenth centuries) British earthquakes seemed engaged on a vendetta against ecclesiastical architecture. (Or maybe these were simply the largest and most fragile buildings?)

It began with an earthquake on 15 April 1185. Declared 'the worst ever known in England', it was felt throughout the country but especially in the north. 'Stones were split', 'stone houses were thrown down', and the Norman Lincoln Cathedral was 'split from top to bottom'.

Damage to the cathedral in the 1185 earthquake seems to have been extensive. Little of the original building remains—only the lower part of the west end and two attached towers. 'Split from top to bottom' implies damage from slow shaking in a larger-sized earthquake.

The next earthquake to maul a cathedral was at St David's in the south-west corner of Wales, where on 20 February 1247 'stones were split'.[2] The following year it was Wells Cathedral under attack. Matthew Paris reports, 'walls of buildings were burst asunder, the stones were torn from their places, and gaps appeared in the ruined walls.'

The 11 September 1275 earthquake was reported to have destroyed the wooden tower of the church of St Michael located on 158-metre-high Glastonbury Tor. (There is even a memorial notice commemorating the earthquake attached to the rebuilt fourteenth-century stone tower of St Michael's.)

A century later on 21 May 1382 we have the largest British earthquake from the later medieval period. These were turbulent times, a year after the brutally suppressed 'Peasants' Revolt', and only thirty years after the Black Death had killed up to half the population. A council of the established church in England was meeting in Blackfriars to discuss how to respond to some of the foreshocks of the Reformation (and thereafter was known as 'the Earthquake Synod').

For the 1382 earthquake we not only have descriptions and locations of damage but even some surviving information about the cost of the repairs. At Hollingbourne in Kent, repairs to the great house and church cost 48 shillings and twopence. (In 1375, 25 shillings would buy 100 feet of ashlar stone.) The archbishop ordered financial relief for the rector after the church's chancel suffered 'grave ruin'. Today the walls of the church are unusually thick and mix in older building rubble.[3] At Canterbury Cathedral many church windows were broken, the bell-tower collapsed, and the monastery and infirmary chapel were seriously damaged. Work on the new nave did not restart until a decade later. In London, both St Paul's Cathedral and Westminster Abbey were damaged and probably also the bell tower of St Bartholomew the Great. Presumed earthquake-related rebuilding is mentioned at the church of St Mary's, Maidstone and Saltwood Castle near Hythe.

The 1382 earthquake also caused damage to buildings on the French side of the Straits of Dover. The moment magnitude is estimated as Mw 5.8. The main shock was followed by several strong aftershocks.

Almost 200 years after the 1382 earthquake there was another, almost as large, probably located offshore between Kent and Belgium. The shock in the early evening of 6 April 1580 was felt across much of England, north-west France, and Belgium. The magnitude is estimated as Mw 5.5. The strongest damage was in Kent. The tower of the parish church of St Peter's in Broadstairs was cracked all the way down—the repairs can still be seen. Two churches in Sandwich were damaged, and the church of St Peter and St Paul in Sutton near Dover was thrown down.

Across the Channel in Calais, houses were ruined and the walls of the city and a watchtower collapsed. The church was also damaged at Boulogne. In Oudenaarde, Belgium, people were killed by falling stones and tiles, and there was chimney damage as far as Brussels.

Damage also extended to London, where dislodged stones from the roofs of Christchurch in Newgate killed two children.

William Shakespeare was just coming up to his sixteenth birthday when the earthquake hit—he began writing plays in the 1590s. In *Romeo and Juliet* Shakespeare gets Juliet's Nurse to recall an earthquake eleven years before. Having a remembered earthquake gave the play an Italian identity. At the same time the reference would not have been lost on the first 1595 London audiences who would have recalled where they were on that spring early evening, between 5 and 6 p.m.

The shaking had been widely felt inside houses, including by the Queen. The reference also triangulates back to Italy. The city of Ferrara was largely destroyed by an earthquake in 1570, which was strongly felt in both the nearby cities of Verona and Mantua where the scenes of *Romeo and Juliet* are set. In an era before newspapers, the 1580 earthquake triggered pamphlets written, published, and sold to satisfy the popular concern and to claim this was God's warning to sinners.

★

What do British earthquakes reveal about today's tectonics (Fig. 64)? Do the faults which are currently active include those that were switched on through one or another period during the past 66 million years?

We can start with the hive of activity around the south-east corner of Kent. On 28 April 2007, a shallow Mw 4.0 earthquake, located 5 kilometres beneath Folkestone, damaged 1,326 properties, more than any other British earthquake since 1931. Many of the older, late Victorian to Edwardian properties suffered fallen chimneys and cracked walls. In some streets two out of three houses were damaged. The local council instituted emergency measures for public safety and evacuated some residents.[4]

Typically for a shallow earthquake, the felt area of the 2007 shock was small relative to the magnitude.

The two largest of all Kent earthquakes, in 1382 and 1580, probably had offshore epicentres, and lay on a zone of seismic activity extending east into Belgium. In the build-up to war, on 11 June 1938, an Mw 5.0 earthquake close to Oudenaarde damaged 17,500 chimneys in Belgium and 1,400 in France.[5]

Through history, shallow-sourced earthquakes have popped off across the Weald and Wessex basins. On 22 May 1201 a series of earthquakes occurred at Montacute in Somerset. On 25 May 1551 an earthquake described as 'terrible' was strongly felt in Dorking. On 9 June 1761, a shock at Shaftesbury caused the end of a cottage to

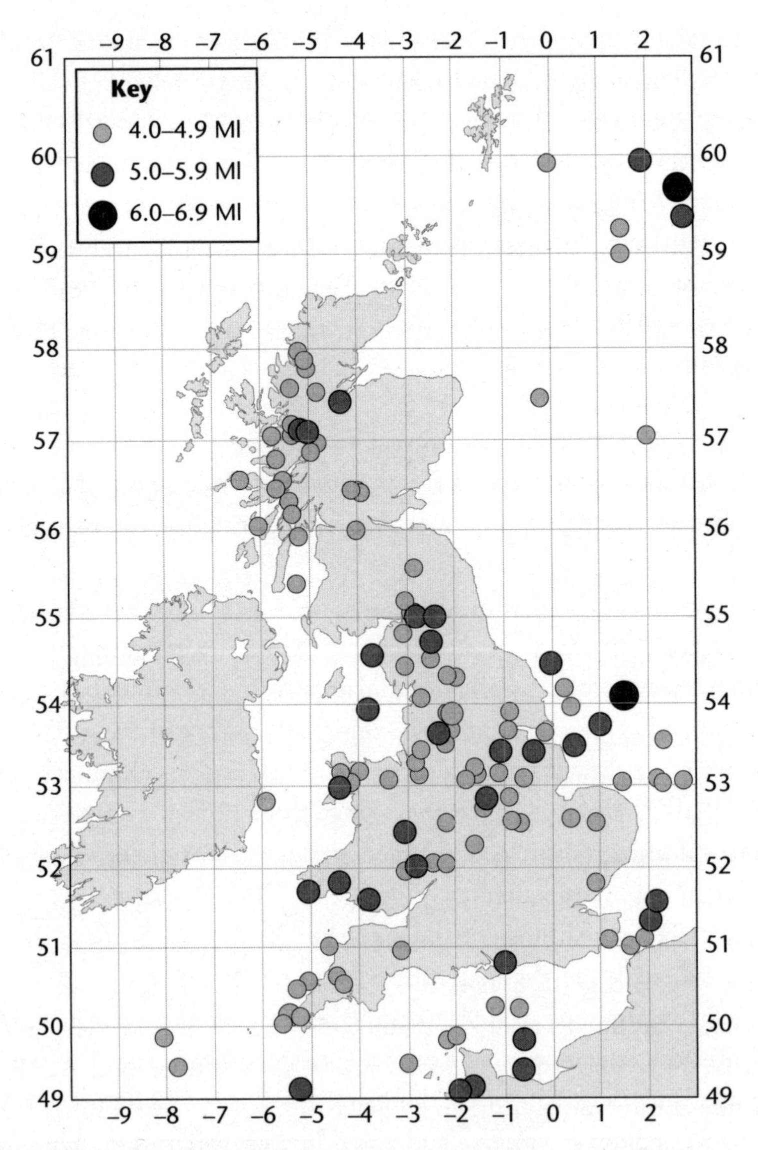

Figure 64. 'Quake Britain'—earthquakes through history

collapse while new springs 'as black as ink' appeared. On the south side of the Weald anticline, to the north of Chichester, on 18 September 1833 one of a sequence of earthquakes triggered a rockfall killing a quarryman. In another shock 'innumerable' chimneys fell and windows broke.

In 1750 earthquakes came to London. The first shock on 8 February was strongest in the East End of the city, at Limehouse and Poplar, where chimneys fell, as also in Leadenhall Street and Southwark. The shaking was felt in towns all around.

Exactly one lunar month later, on 8 March, there was a second stronger shock, spilling tiles off roofs and damaging chimneys as far as Croydon. Part of a house fell in Old Street and two uninhabited houses collapsed in Whitechapel. The top of one of the piers of Westminster Abbey tumbled down. The earthquake was felt more widely than the first.

The tabloids of the age published a prophecy from an apocalyptic guardsman that these shocks were portents of a truly devastating shock destined for another lunar month later on 5 April. This prophecy generated a panic. Those with money left the city, whereas those too wise, or poor, to leave held defiant parties in the parks, after preparing special sleeping suits. April 5th proved quiet, leaving all who fled to cover their embarrassment. However, in this same year of earthquakes, there were seven shocks felt across England.

The devastating urban catastrophe followed five years later, not in London but in Lisbon.

For the Victorian era, the 'great British earthquake' occurred on 22 April 1884 with an epicentre a few kilometres south of Colchester, Essex near the village of Peldon. The earthquake shaking caused extensive damage to houses and churches, unusual for Britain. These damages declined rapidly with distance, indicating a shallow earthquake source, perhaps less than 5 kilometres underground. A bronze plaque commemorates the earthquake in the restored church at Little Wigborough (only the second earthquake memorial in Britain). The local water table, monitored by a brewery, rose after the earthquake, responding to the shallow fault movement.

The largest earthquake in the wider British region, at least since 1382, occurred early morning on 7 June 1931 beneath the Dogger Bank, 150 kilometres offshore from the coast of east Yorkshire. The moment magnitude is assessed as 5.8 Mw. The shaking was felt all around the North Sea. Along the east coast of England, chimneys toppled in coastal towns from Whitby, Hull, Beverley, and Bridlington down to Great Yarmouth. Worst affected was the nearest coastal town at Filey. There was an extensive rockfall from the chalk cliffs at Flamborough Head.

Across the southern half of Wales, Swansea has been repeatedly in the earthquake front line in 1727, 1775, and 1906. The Welsh borders have also proved fertile earthquake territory. On 17 December 1896 an earthquake toppled more than 200 chimneys in Hereford, with the greatest concentration of damage 6 kilometres east of the city. Late night on 2 April 1990, there was an Ml 5.1 (Mw 4.7?) earthquake close to the town of Clun. The depth was 15 kilometres underground, so although the shaking was widely felt, the shock caused only scattered gable, roof, and chimney damage in this rural area—a very 'British' earthquake.

In the vicinity of north Wales, earthquakes seem to be concentrated around the Llŷn Peninsula to the west of Snowdonia. In the early hours of 9 November in 1852, a shock was strongly felt in Dublin, although the epicentre was probably closer to Llŷn. On 19 July 1984 the largest earthquake known from the region, with an Mw of 5.0, occurred 23 kilometres underground, unusually deep and directly beneath the centre of the Llŷn Peninsula. This cluster[6] suggests the Miocene tectonic legacy still drives seismic activity today.

Earthquake activity in Scotland is concentrated out to the west of the Highlands. Scottish earthquakes have a tendency to arrive in swarms, with repeated events of similar, modest size.

The town of Comrie on the Highland Boundary Fault was once claimed in the local press to have suffered more earthquakes than anywhere on Earth. One swarm of shocks started in 1788, taking thirteen years to build to a climax in 1801. Another swarm was initiated by the largest earthquake of them all on 23 October 1839 when many houses were damaged. A committee was set up by the British Association for the Advancement of Science to study the phenomenon and a garden-shed-sized stone 'earthquake house' was built in a field near the centre of town for performing scientific experiments in earthquake measurement. The only experiment actually installed consisted of solid metal cylinders of different heights and diameters, standing on a flat surface, intended to fall in the direction away from the earthquake source. A single cylinder toppled in one of the subsequent earthquakes. The swarm continued for seven years, with long gaps followed by a spate of tremors. Comrie's benign earthquakes became something of a tourist attraction.

Other earthquake swarms have buzzed through the Ochils south of Comrie as well as rattled Glendevon in 1981 when the coping stones across a local dam were found to have popped up. The town of Inverness suffered swarms of tremors culminating in the largest known earthquake in Scotland (assessed as Ml 5.2) on 13 August 1816. The earthquake tumbled chimney pots, but avoided casualties, arriving at 10.45 p.m. when most God-fearing Highlanders were already in their beds, from whence they fled in their nightclothes.

★

The fastest waves that radiate through the Earth away from the sudden fault rupture are known as 'p waves', or pressure waves. They are like sound waves in air, and involve movement towards (compression) and away (dilation) from the direction in which the waves are heading. (The next waves to arrive are the slower 'S waves', or shear waves, that vibrate from side to side.)

At the point where the fault starts to break, the first impulse creates a positive pressure wave along the orientation of the fault rupture, in the direction of movement. As the two sides of the fault start to move in opposite directions, two positive pressure waves head away, 180 degrees apart, separated by the fault. On each side of the fault, opposite the wave of compression, the pressure in the rock is being relieved. This propels a negative wave of dilation (pressure reduction).

At each recorder the sign of the initial pressure wave is automatically read, whether it was forwards or backwards, compression or dilation. Where there are enough recorders in different directions, and at different distances, around the fault rupture, the patterns of compression or dilation make it possible to identify the orientation of the fault-source and the direction in which it broke. This can reveal what is driving today's tectonics. Where there is ambiguity about the fault orientation, the location of aftershocks can resolve the question.

It turns out that recent British earthquakes are principally being caused by horizontal or strike-slip fault movements.

For example, the 2 April 1990 magnitude Ml 5.1 earthquake at Clun involved sinistral horizontal movement on a N–S fault.[7] Likewise the 14-kilometre-deep Ml 4.7 earthquake at Dudley, in the English Midlands on 22 September 2002.[8]

The 1984 earthquake beneath the Llŷn Peninsula of north Wales involved dextral displacement on a WNW fault, and the 2006 Folkestone earthquake was also strike-slip.[9]

The pattern of fault movements across England and Wales seems consistent. The crust is being squeezed NW–SE, as has been the case since the Miocene. However, today the crust is at the same time expanding NE–SW.[10]

Meanwhile, earthquakes in west Scotland reveal a distinct stress field with the maximum compression direction oriented north–south. This suggests a different cause, potentially from the unloading of the later ('Younger Dryas') ice sheet.

Earthquakes reveal where the crust is actively deforming.[11] 'Quake Britain', as we could call this territory (see Fig. 64), has a different outline to Geographic Britain defined by the current coastline. The western coastline of Geographic Britain runs close to the western edge of Quake Britain. However, Ireland is almost completely aseismic—in earthquake terms it is dead, undeforming.

To the south, Quake Britain continues without interruption across the English Channel into northern France and across the Straits of Dover into Belgium.

Quake Britain also extends offshore from Lincolnshire and Humberside into the western part of the southern North Sea.

North-east England and eastern Scotland lie outside Quake Britain, as does the western half of the northern North Sea. Quake Scotland is more or less confined to the Highlands, and does not continue far into the Hebrides.

Across Quake Britain the density of earthquakes has not been uniform. There has been a nest of earthquakes around the Dover Straits and another around the Llŷn Peninsula in north-west Wales. A band of activity runs through southern Wales into the borderlands with England.

Across Quake Britain earthquakes are not concentrated on a single fault or set of faults like a plate boundary, rather they are being driven across a large number of small faults across a region. Something is driving this deformation, and the earthquakes that it generates, most likely related to the uplift of the past 2.5 million years.[12] The chatter of earthquakes reveals low-level tectonic activity, with earthquakes firing off where some of the big faults broke in the past. And we have simply been lucky not to have recently experienced a 1382 earthquake or a 1931 earthquake with an epicentre on land, from which there could be a lot of damage and casualties. There is no way of knowing how much bigger an earthquake could be. The only comfort is: the larger the shock, the rarer it is.

★

Before seismic recorders we only know about earthquakes because enough literate people felt the shaking (or heard about the shaking) and chose to write about their experience.

Before people could write they still experienced the earthquakes, but had no ability to leave a report.

Before people, the Earth still shook. Like the tree in the forest that falls and no one hears, over the ages there have been millions of earthquakes, witnessed only by the animals and the trees.

Unwitnessed, but maybe not without leaving some impact.

In the western United States, geologists map 'precarious rocks' poised alarmingly on some rock platform to show where there has *not* been strong shaking. On the same principle I was once commissioned to check out the surviving, fragile, Iron Age roundhouse brochs of north-west Scotland. Those still standing could not have experienced strong shaking over the past 2,000 years. (Many of them, for one or another unknown reason, have collapsed.)

And beyond the consequences of shaking, is there evidence of a prehistoric fault rupture big enough to break surface?

A few tantalizing clues suggest some recent fault displacement cutting right through to the surface, at Beeston and Skelding Hill, in Norfolk. There is even a suggestion of recent fault movements[13] beneath Farringdon in central London.[14] The Hackney Gravel, a 250,000-year-old Thames river terrace deposit, rests on a surface at 13 metres elevation to the west that drops down to 11 metres to the east.

Fault movements are motivated by the stresses accumulated in the crust. We can even use tectonic arguments to show that some suspect evidence can't be recent faulting. A series of NE-trending troughs identified from boreholes at the base of the Red Crag in central East Anglia have been argued to show evidence of recent normal faulting,[15] but the orientation and extension direction are incompatible with the prevailing stress field. Instead these Suffolk chalk troughs must have been carved by deep scouring from tidal or river currents (or are re-eroded older fault scarps).[16]

Similarly, the proposal that the Sticklepath Fault moved sinistrally through the Cenozoic[17] is incompatible with the prevailing and very consistent stress field. You can't argue with the physics of tectonics.

*

Away from active volcanoes, hot springs may also record past earthquakes. The Earth's temperature rises 15–30 degrees Celsius for each kilometre underground. To bring deep-sourced hot water up to the surface requires an open channel, or 'chimney' through the rock, accompanied by a strong flow of water. If the flow was squeezed through hundreds of tiny cracks, the water would cool to the ambient rock temperature as it rose. If the water infiltrated its way upwards over thousands of years, or passed through wide karst limestone caverns, it would not arrive at the surface hot.

When the hot spring originally formed, the chimney must have appeared suddenly. New hot springs have been observed to emerge after significant earthquakes, and many thermal springs are situated on faults.

The relationship between the chimney and the fault movement is subtle. If the aperture was too big, like a pull-apart basin, the sides could collapse and block the channel. So the ideal is a small displacement on a fault that goes deep, reaching down to some 'artesian' hot-water aquifer and then, accompanying the earthquake, you have a new thermal spring. The thermal spring at Taff's Well near Cardiff even arrives at the surface up an outcropping fault.

Over time the chimney gets constricted by minerals, crystallizing as the cooling water moves towards the surface. As the aperture of the chimney shrinks, the water slows and cools faster until mineralization blocks the whole channel. Seen in outcrop in a cliff or quarry, veins of quartz and calcite fill the cracks around faults. The thermal spring has a life span: born in fault displacement and staunched by 'mineral sclerosis'.

Across Britain there are two zones of thermal springs (Fig. 65). By far the most impressive is at the city of Bath, founded and named by the

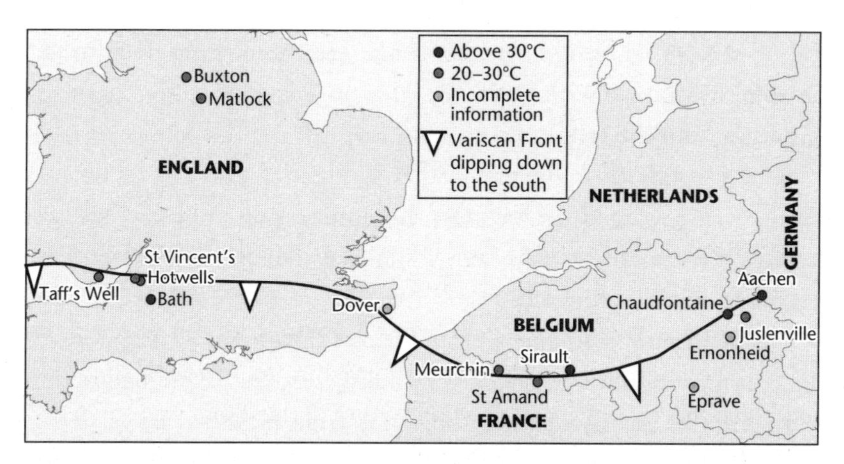

Figure 65. The land of thermal springs, from Wales to the Germany/Belgium border

Romans, where there are three thermal springs (King's, Hetling, and Cross Bath) in an area the size of three tennis courts in the flood plain of the River Avon. Fifteen litres of water emerge each second, at a temperature of 45°C and an artesian head 9 metres above ground level. The water is hot because the flow is strong. At depth the chimney up which the water flows is likely to be fault-defined.[18] Sixty metres below the flood plain, beneath a layer of flat-lying Triassic sediments, the Carboniferous limestone strata dip steeply and the geological structure is faulted and disturbed.[19] Geochemistry reveals the water started at temperatures between 64 and 96°C, 4 kilometres underground.

If we knew when the Bath thermal spring first formed, from dating accompanying mineralization, we could identify when were the likely fault movement and accompanying earthquake that brought it into existence.

The River Avon runs to the WNW from Bath. The Avon Gorge, connecting Bristol to the sea, may follow a line of WNW fractures. Along this line there are two thermal springs: one at Hotwells at the start of the Avon Gorge and a second, 3 kilometres into the gorge on the northern side. The Hot Well in Bristol flows at 4 litres a second, but two-thirds of this flow is groundwater from the limestone through which the water emerges[20] and so the temperature is only 24°C. In the eighteenth century, attempting to rival Bath, the warm water was pumped up to a fashionable spa hotel. The spa is long demolished, unable to compete with Bath's superior product, and today the spring emerges from the river mud close to low tide.

The second warm spring farther along the gorge (St Vincent's Spring) also gained an eighteenth-century pumphouse. This was removed for a railway and later a highway, but more recently fed a roadside drinking fountain.

The Bath and Bristol springs have formed close to the Variscan Front,[21] in tectonic situations comparable to the Wessex and Weald basins, so it would not be surprising if the underlying faults remained active.

Although it does not reach to the surface, thermal water was encountered while excavating a shaft in search of coal seams at

Shakespeare Cliff near Dover. On 6 March 1897, digging about 90 metres below the surface, a sudden outburst of warm water flooded the shaft, drowning eight miners.[22] No one seems to have measured the temperature, so we don't know how warm it was.

Hot water was also encountered at several locations deep in Cornish tin mines. At the Crown Point Mine, more than 400 metres below ground level, the water arriving up a borehole was apparently 'hot enough to boil an egg'.[23]

Farther to the east there are thermal springs along the Variscan Front through Belgium. The word 'spa' comes from the name of a Belgian town, south of Liège, famous for the health-giving properties of its springs. There is the Stambruges thermal spring near Mons[24] south-west of Brussels, but the hottest spring is found alongside NW-trending faults at Liège, itself the site of a shallow and widely damaging earthquake in 1983.

Fifty kilometres farther east at the ancient German city of Aachen, the Variscan thrusts are cut by active NW–SE-trending normal faults. Aachen has been a long-term earthquake centre: the largest documented shock was in 1756. Thirty hot springs emerge in and around the city at the Variscan Front with temperatures as high as 74°C, and a flow of 3.5 million litres each day, compared with Bath's 1.3 million litres.. As at the city of Bath, the original bathhouses were constructed by the Romans. The endless supply of hot baths was what brought the Emperor Charlemagne to live here.

The only other warm springs in Britain are located around the Derbyshire Peak District where ten 'thermal' (i.e. above ambient 9°C groundwater) springs emerge from valley floors.[25] At Buxton the water arrives at 27°C, from fissures in Carboniferous limestone, whereas at Matlock the temperature is 20°C.[26]

All the hot spring locations in Britain lie beyond the maximum extent of the last ice sheet. The pressurized water beneath an ice sheet would likely reverse and kill a thermal spring (there are none in Scotland or Scandinavia).

★

Is the Sticklepath Fault active? Could it launch a significant earthquake?

We know some NW–SE faults are active. A swarm of earthquakes followed this alignment beneath Manchester.[27]

On 29 September 1858 an earthquake was felt along the north edge of Dartmoor, 'most strongly at Sticklepath and the neighbouring villages'. The high intensity and small felt radius indicate a shallow source, potentially on one of the strands of the Sticklepath Fault.[28] The Codling Fault looks active. Thirty kilometres north-east of Dublin, the sinuous Lambay Deep, reaching 120 metres water depth, is like a narrow pull-apart basin along the Codling Fault which also presents steep sea-floor fault scarps and microseismic activity.[29] Escaping methane gas has created a line of carbonate mounds, each 5–10 metres high.[30]

Yet the tectonics that once short-circuited between the Norwegian Sea and Mediterranean plate boundaries has moved on. The era of multi-metre Sticklepath fault displacements, and their accompanying shattering earthquakes, is as much prehistory as the dinosaurs.

★

To the south-west of Britain lies the current plate boundary separating the African and Eurasian plates (Fig. 66). The boundary runs east from the mid-Atlantic spreading ridge around the latitude of Lisbon and on into the Mediterranean.[31]

The plate motion skews; to the west, slow opening has created volcano-filled rifts that make up the Azores islands. Farther east, midway to Portugal, there is pure east–west strike-slip displacement along the Gloria Fault. The eastern end broke in a 160-kilometre-long rupture in November 1941, accompanied by an Mw 8.3 earthquake.[32]

Closer to Portugal the plates are colliding obliquely and it becomes impossible to define a narrow boundary.

South-west of Portugal lies the 60-kilometre-wide and 180-kilometre-long NE–SW-oriented 'Gorringe Ridge', which rises from the 5,000-metre-deep sea floor to within 25 metres of sea level (enough to graze the hull of a laden supertanker). Adjusting for latitude, at the sea

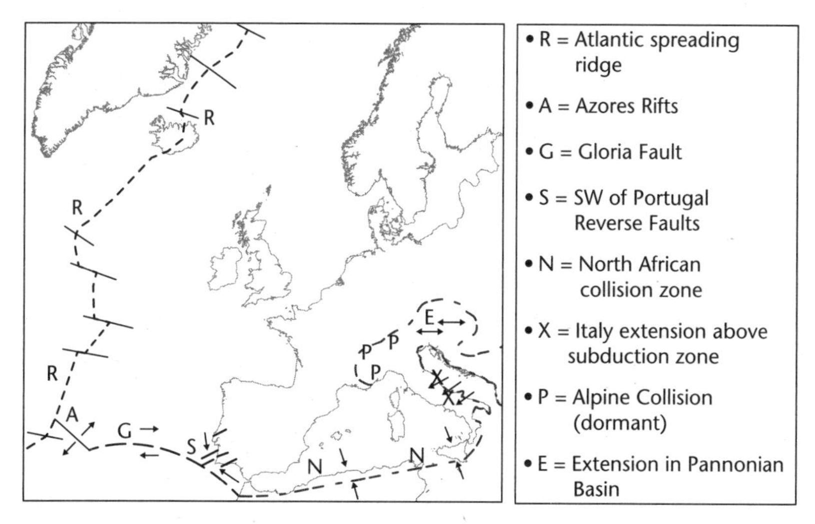

Figure 66. The plate boundaries around Britain today

surface this mountain range exerts the strongest pull of gravity anywhere on Earth. Gorringe Ridge is composed of heavy oceanic crust and mantle rock. Only the most powerful of horizontal tectonic stresses from the Africa–Eurasia collision zone could have forced this massive, dense mountain range to be raised up to the ocean surface.[33]

South of the ridge a 600-kilometre-long ESE-trending strike-slip fault has recently been identified heading for the middle of the coast of Morocco.[34]

Yet all the collision at this plate boundary seems to be absorbed to the north of this fault, as in the 1969 Mw 7.8 earthquake, situated on a NE–SW-trending reverse fault at the foot of the Gorringe Ridge.[35]

Although the 1969 earthquake was big—the largest across Europe since 1900—it was one-twentieth the size of what happened on the first day of November in 1755.

Around ten o'clock, on a sunny morning, at the 30-acre Pibley Pond in Derbyshire, the gamekeeper watched spellbound as the water rose and fell 2 feet and continued oscillating for two hours.[36] At a lake at nearby Dunstall Hall the water rose for several minutes 'like a pyramid' and 'fell down like a waterspout'. At the shores of lakes across northern England and Scotland, the water rose for several minutes and then slowly receded, repeating the strange movement for up to an hour. At Loch Ness the rise in water levels was 'so violent as to threaten some houses'.[37]

The stories were recorded in letters and published in newspapers in Edinburgh and London. A few days later news arrived from Holland and the northern German states of rivers, canals, and moats thrown into oscillation at the same time. After two weeks, reports arrived of churning lakes from Switzerland, Sweden, and Norway. It was like some cosmic shudder. Nobody knew what had caused it. Nothing like this had ever been witnessed before.

It took four weeks for the news to reach London that on the morning of 1 November shaking, tidal waves, and fires had destroyed the city of Lisbon.

This 1755 earthquake was huge—in the top rank of giant earthquakes[38]—approaching magnitude 9: a size only experienced globally on average a few times a century, comparable to the 2004 Boxing Day Indian Ocean earthquake or the 2011 earthquake that devastated the Pacific coast of Japan.

A magnitude 9 earthquake is generated by a fault at least 250 kilometres long with a displacement of 20 metres or more. All other earthquakes of this size over the past 250 years have been situated on subduction zone 'megathrust' plate boundaries.

So where was the fault that ruptured to generate the 1755 'Great Lisbon' earthquake?[39] The timing of the accompanying tsunami shows it was generated not far offshore from the coast of south Portugal, close to the western Algarve, and continuing farther to the south. Tectonicians argue over whether they see signs of a subduction zone forming south-west of Portugal.

And if this riddle is not enough, five years later at the end of March 1761, there was another great earthquake, which also shook the whole Atlantic margin of south-west Europe. The event is almost unknown (reports were suppressed in Portugal for fear of provoking alarm). This may be the closest to Britain any really big earthquake has come over the past few centuries. This earthquake was not quite as large as that in 1755, but at least magnitude 8.5–8.8 in size. In Lisbon[40] violent shaking lasted five minutes, as compared with six minutes in 1755.

The area that experienced the shaking was vast. The vibrations were strong enough to cause the collapse of a few fragile buildings in towns and villages across Portugal, from Lisbon and Setúbal in the south and into northern Portugal. At the port of A Coruña,[41] in north-west Spain, the earthquake lasted 'some minutes' and 'caused houses to slide a few feet down the slope'. At Funchal on the island of Madeira, 1,000 kilometres south-west of the Portuguese mainland,[42] the walls of several buildings were damaged, while on the eastern coast rockfalls killed some sailors and demolished a church. The shaking was felt from Agadir on the southern coast of Morocco, to Cork in Ireland, over a span of 2,400 kilometres.

The earthquake was also large enough to generate powerful slow vibrations, beyond the area in which the shaking was felt. These were not as widespread as in 1755 but were still impressive, swinging chandeliers in churches and synagogues in Amsterdam, and sloshing the water in Loch Ness for 45 minutes.[43] (The local Highlanders assumed a monster had once again disturbed the lake.[44])

The displacement of the sea floor generated a tsunami. At Lisbon the water ebbed and flowed by up to 2.5 metres, starting 75 minutes after the shaking. The tsunami was 1.8 metres in Cornwall and 1.2 metres in southern Ireland, but fortunately arrived at low tide. At Portorico on the Azores island of Faial, all the lighters and fishing boats hauled out of reach of the sea were carried away and broke to pieces on the rocks. On the other side of the Atlantic, at 4.30 p.m. local time, the sea at Barbados suddenly retired and then rose again by 1.2 metres, continuing to oscillate for hours.

All these phenomena confirm this was a huge earthquake. The 75 minutes it took the tsunami to arrive at Lisbon, and the absence of any severe damage, tell us the source was farther offshore than in 1755. Perhaps the most useful observers would therefore have been on ships?

Around midday on 31 March 1761 a number of ships sailing between Portugal and Britain reported violent shaking.[45] One captain mentioned the spindle of the compass flying off, another that the wooden balks of the ship opened so that water flooded in, causing sailors to panic and throw objects overboard to save their ship. The captain of a third vessel believed they had struck a rock. The ships that felt the shock strongly included the British Expedition Packet Boat, tossed about for four minutes while located offshore from the Rock of Lisbon (the outermost cape on the north side of the Tagus estuary). Farther north, several British vessels were in convoy, keeping a wary distance from the enemy Spanish coast.

The ships badly shaken by the 1761 earthquake were scattered from Lisbon northwards for at least 500 kilometres (Fig. 67). We can compare this pattern with the locations of the vessels that reported strong shaking from the 1755 earthquake—sailing to the south of the Rock of Lisbon, where other evidence suggests the earthquake was located. We also have reports from ships in the vicinity of the 1969 earthquake source, 150 kilometres south-west of Portugal. One tanker directly above the sea-floor fault rupture suffered such damage it had to return to dry dock. Two other vessels in the region also strongly felt the seaquake.

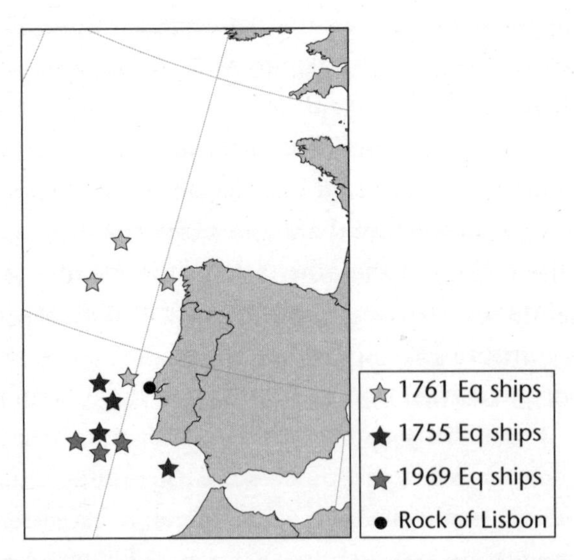

Figure 67. Locations of the ships strongly shaken by the great 1755, 1761, and 1969 'seaquakes'

From the ship reports we would place the 1761 fault rupture 100–200 kilometres to the west of north-west Spain.[46]

In the vicinity of Portugal, the African and Eurasian plates are colliding at around 4 millimetres each year. For comparison, the boundary between the North American and Pacific plates through California moves at 50 millimetres per year: more than ten times faster.

Below the sea floor west of Portugal the biggest earthquakes, with displacement of tens of metres, like those in 1755 or 1761, take thousands of years to repeat. This contrasts with the hundreds of years it takes for earthquakes to repeat along the San Andreas Fault in California. Along the 'fast' (80 millimetres per year) plate boundary offshore from central Chile, Darwin's 1835 earthquake more or less repeated in 2010. For the ocean margin of Portugal and Spain, that makes it a particular challenge to assess the hazard from big earthquakes. We can expect future shocks with no historical precedent.

From the middle Miocene until recent times, Spain was squeezed by strong NNW-directed crustal compression as it absorbed some of the Africa–Eurasia collision,[47] raising a set of internal ENE-trending mountain ranges. This shortening seems to have ended and although there is some limited earthquake activity around the Pyrenees,[48] GPS survey measurements over ten years have not detected active tectonics.

In earthquake terms the western Alps have also gone quiet. Within the past 2.5 million years, the Upper Rhine Graben has returned to extension, with displacement on some boundary faults close to Heidelberg accumulating up to 300 metres of Pleistocene deposits.[49] At the southern end of the graben, the largest (Mw 6.5?) earthquake north of the Alps in the past thousand years devastated Basel by shaking and fire on 18 October 1356, destroying churches and castles out to a radius of 30 kilometres.

12

The rocky road

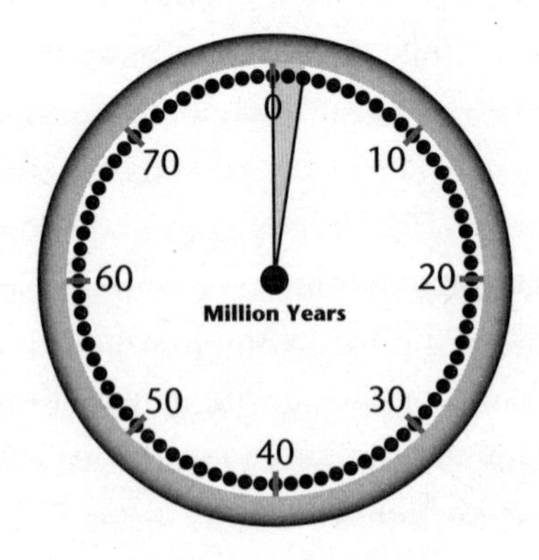

Lewis Fry Richardson was a mathematician and Quaker who refused to fight in the First World War. Ever after, on account of his defiant pacifism, he was unable to obtain an academic position. Yet he wanted to apply his mathematics to forging peace. One of his obsessions became finding the mathematical equation for measuring 'hostility' between countries. Maybe, he thought, there could be some dependence on the length of their mutual borders: the longer the frontier, the greater the source of aggravation.

However, when he sought to find measurements on the length of specific frontiers he encountered a problem. Adjoining countries might have very different perspectives. The Spanish declared their border with Portugal ran for 987 kilometres, whereas the Portuguese asserted this same frontier was 20 per cent longer at 1,214 kilometres.

They could not both be correct. Or could they? Richardson dug deeper into how the lengths had been measured.

Imagine using a pair of dividers on a map, or a measuring pole swinging from one point to the next along the border. The larger the dividers, the more the fine detail along the frontier gets missed and the shorter the apparent distance covered. There is no 'correct' length.

Richardson plotted the logarithm of the size of the measuring device against the measured length of that specific border. As the size of the measuring device is reduced, the length increases.

Mainland Britain has no frontier except with the sea. Taking the west coast of Britain, Richardson found one of the best examples yet of this pattern. The length of the coastline consistently increased as the measuring device was reduced from 1,000 kilometres down to 1 kilometre.

In the 1960s Richardson's work was expanded by the Polish-born mathematician Benoit Mandelbrot who was studying the behaviour of financial markets while working for IBM. The changes in the value of some commodity—the size of the movement from one day to the next—were just like processes in the natural world. 'Clouds are not spheres,' he wrote, 'mountains are not cones, coastlines are not circles, and bark is not smooth, nor does lightning travel in a straight line.' He termed these structured examples of randomness 'fractals'. For a true fractal there is a linear relationship between the logarithm of the sampling size and the log length (or number or volume). The gradient of this linear log-log relationship is said to be its 'dimensionality'.

In 1967 Mandelbrot published a paper in *Science* magazine titled 'How Long Is the Coast of Britain? Statistical Self-Similarity and Fractional Dimension'.[1] Thanks to Richardson the British coastline had become a fractal icon (Fig. 68).

With a true fractal, you have 'scale invariance'. As the frame shrinks, the complexity stays the same. The steam from a kettle is like a patch of fog is like a cloud. The twigs at the end of a branch are arranged like the branches themselves. A fragment of rock resembles a mountain.

Make a map of all the big fractures that characterize a broad fault zone exposed in a cliff. The map resembles the pattern of fractures seen in a boulder from the fault zone, or even the tiny fractures seen

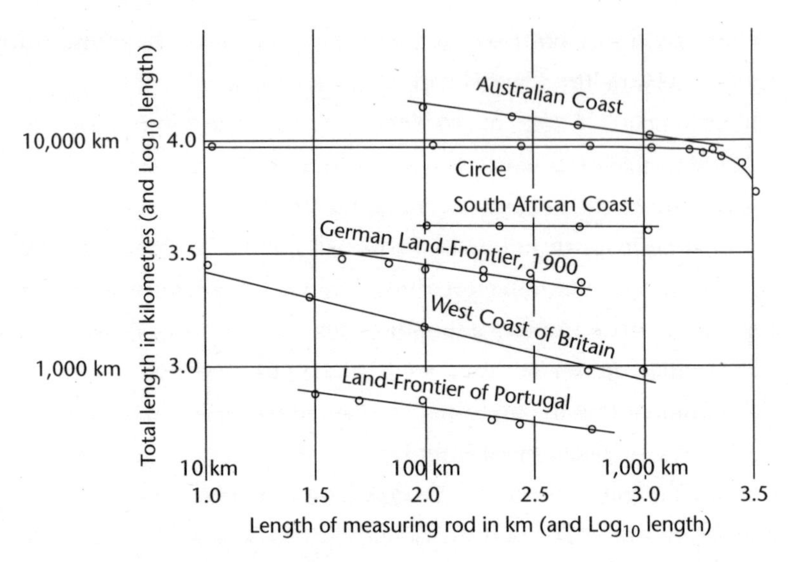

Figure 68. The west coast of Britain: a fractal 'icon'

in a thin section of the rock viewed down a microscope.[2] The fracturing is 'scale invariant'. Without a scale bar we would not know the size. (This is why geologists like a 'person for scale' in their photographs of the rocky landscape.)

Earthquakes also show fractal behaviour. Richter's magnitude was derived from the logarithm of the amplitude of the shaking. Plot the cumulative number of earthquakes up to some magnitude and we have the linear 'Gutenberg–Richter relation'. The population of earthquakes relative to their sizes is a fractal (i.e. there might be ten magnitude 3 earthquakes for every magnitude 4, and ten magnitude 4s for every magnitude 5), just as the population of fractures of different sizes is a fractal.

One question that neither Richardson nor Mandelbrot thought to ask: why does the west coast of Britain provide such a fine fractal? Richardson had already tested other coastlines—he gave the example of South Africa—where the length did not change appreciably with the size of the measurement instrument. Many other coastlines, such as western Norway (too many similarly dimensioned fjords), or even the eastern coast of Britain with its long sections of smooth beach coastline, do not make convincing fractals. In fact, it soon becomes clear why Mandelbrot, borrowing from Richardson, framed his paper around the coast of Britain: it is hard to find a better example.

What has made the west coast of Britain so irregular on all scales?

First, there is the tectonics. The west coast is made up of ruined volcanoes, faults, and pull-apart basins. Faulting has juxtaposed hard rock against soft sediments, to be exploited by the waves, carving complex coastlines, whether the coves and headlands of Cornwall or the peninsulas and sea lochs of the western Highlands.

Next, there is the consequence of lowered Ice Age sea levels leading to numerous estuaries: the drowned 'ria' valleys of rivers that cut down to reach a far lower sea level. We have estuaries because sea level has risen by more than 100 metres in 20,000 years.

Britain's maritime power owed a lot to these estuaries, which provided perfect deep-water sheltered anchorages: think of Portsmouth, Plymouth, or the Clyde. A drowned river valley is the making of the city of London. The river on its own would merit less than 10 per cent of the width and would scarcely be navigable. The largest estuaries, along with the complex coastline, have amplified the tides. Boats could move into and out of the ports of London or Bristol using only the tides, without requiring any exertion of energy apart from hauling and dropping anchor. An enemy boat arriving on the tide could not escape until the tide turned.

The third ingredient to the fractal coastline is the lack of available sediment to infill these estuaries. Apart from a few locations, as at Harlech in north Wales, blown sand has not arrived in sufficient quantities to smooth the west British coastline with a ridge of sand dunes, as along the coast of Holland or around sections of the east coast of England. Golf, with its dune bunkers and glacial till undulations, is an east coast of Scotland game. Long sections of the east coast of England are artificial. The Norfolk Broads were formed by uncontrolled medieval peat digging, after the local forests had been ravaged for fuel and timber. As in Holland, the sea moved in to flood the pits. Without the coastal embankments and flood protection around the Wash, high tide would turn Ely back into an island and king tides would reach Cambridge, as was the case before the twelfth century.

A high tidal range flushes the estuaries and keeps them open. If the river carried more sediment, the estuary would turn into a delta. Even with the rapid rise in sea level ending only 6,000 years ago, the rivers Rhone and Rhine have been able to rebuild their deltas from all the sediment eroding off the glaciated Alps. Meanwhile the sediment-starved Seine and Loire, like the Thames and Severn, sustain their estuaries. There are no significant deltas around the coast of Britain.

Uniquely among fractals, this one can actually be 'experienced' in all its irregularity and diversion, its inlets and headlands, through pacing more than 2,000 kilometres of the south-west England and Wales coastal paths.

*

Sometimes in a pile of strata the younger sediments are more lithified, more resistant to erosion, than the older layers of clay and sand beneath. As an anticline grows through repeated underlying fault movements, this hard layer of limestone or sandstone becomes a raised whaleback ridge. Stream erosion eventually penetrates the carapace, exposing and eroding the weaker sediments at the heart of the anticline. Then the topography reverses. The original culmination of the anticline becomes the most deeply eroded. The hard layer remains all around the rim while the centre is a low-lying hollow.

There is no word in English for this configuration. In French it is known as a 'boutonnière' or buttonhole. There is a classic boutonnière along the great anticline that follows the NW-trending Pays de Bray Fault as it cuts through the Paris Basin en route for the Isle of Wight. Easily eroded Jurassic clays are exposed in the centre, surrounded by steep inward-facing escarpments of chalk. Topography has given the community who live inside the buttonhole physical and cultural isolation.

In the Negev Desert of southern Israel the eroded anticline configuration is termed an 'erosion crater' and given the Hebrew name

'Makhtesh', meaning 'mortar grinder'. There are four such 'craters' in the Negev, each originally an anticline. The largest, Makhtesh Ramon, is 40 kilometres long and up to 10 kilometres wide, eroded to a depth of 500 metres. The original harder layers in the roof of the anticline are exposed all round the rim, but inside the crater are the underlying weaker sediments.

The Israeli Makhtesh is in a desert: the rocks are pigmented in bold reds and ochres, and there is almost no habitation. The Weald Basin is a much greener and more populated Makhtesh. The more resistant Chalk formation and underlying Greensands provide the rim of the North and South Downs (Fig. 69). The interior of the anticline is low-lying, exposing the weaker clays and sands. However, in the eastern axis of the anticline, the High Weald is raised because the oldest of all exposures are harder sandstones. Across the Weald Basin the layers are laid out like a great reptile eye. The dimensions are bigger than in the Negev. Makhtesh Weald, we could call it, is almost 65 kilometres wide and 150 kilometres long, continuing across the channel into the Jurassic coast of the Boulonnais.

We are once again going to encounter Charles Darwin. 'Makhtesh Weald' was to play a key role in his arguments around the role of natural selection in the origin of new species.

Darwin only had to walk 5 kilometres south from his house at Downe, in Kent, to reach 251-metre Betsom's Hill, on the northern chalk lip of the great erosion crater.

Darwin and his wife Emma were frequent visitors to the house of his sister (who was married to his own wife's brother), at Leigh Hill Place, a picturesque 40-kilometre ride along the northern edge of the Weald Basin. Leigh Hill Place, a gaunt Victorian house with high ceilings and a doll's house profile, was situated on the lip of the hill with spectacular views to the south. On a clear day, the ridges of forested hills, each defined by some more resistant geological stratum, succeed each other like waves in the sea. From the eons of geological time required to unroof and erode this great anticline, Darwin perceived the deep time required to explain the evolution of species:[3]

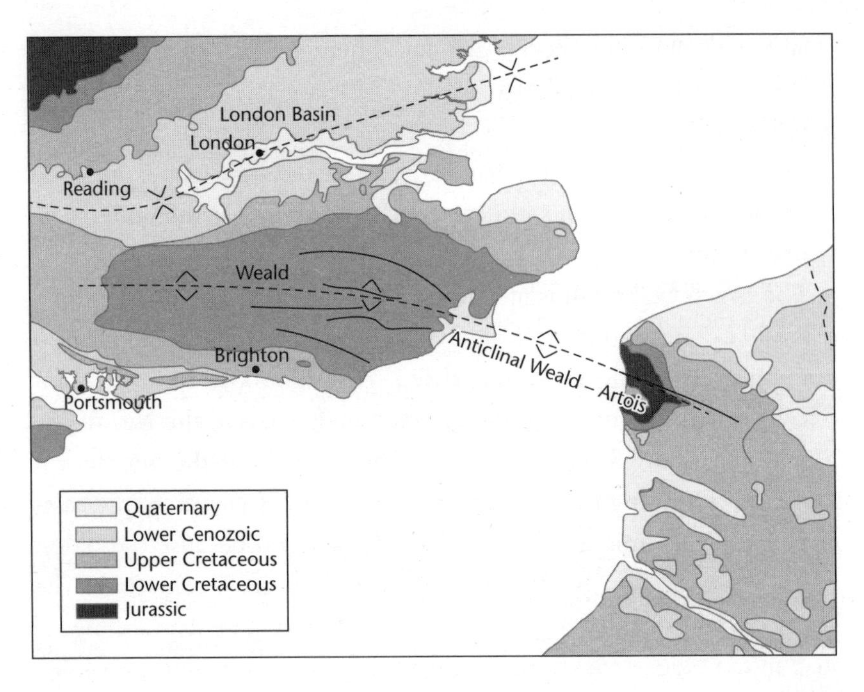

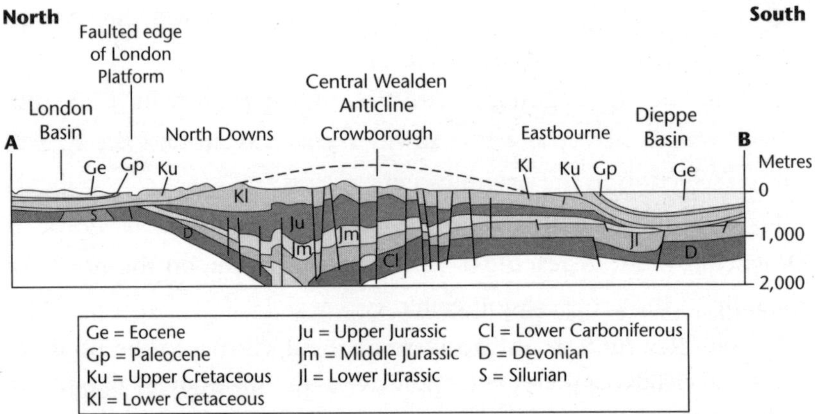

Figure 69. 'Makhtesh' (anticline erosion-crater) Weald

...it is an admirable lesson to stand on an intermediary hilly country and look on the one hand at the North Downs, and on the other hand at the South Downs;...one can safely picture to oneself the great dome of rocks which must have covered up the Weald within so limited a period as since the latter part of the Chalk formation.

...if then we know the rate at which the sea commonly wears away a line of cliff of any great height, we could measure the time requisite

> to have denuded the Weald. . . . we may in order to form some crude notion on the subject, assume that the sea would eat into cliffs 500 feet in height at the rate of one inch a century . . . the same as if we were to assume a cliff one yard in height to be eaten back along a whole line of coast one yard in nearly every twenty two years. And at this rate, on the above data, the denudation of the Weald must have required 306,662,400 years; or say three hundred million years. But perhaps it would be safer to allow two or three inches per century, and this would reduce the number of years to one hundred and fifty or one hundred million years.
>
> The action of fresh water on the gently inclined Wealden district, when upraised, could hardly have been great, but it would somewhat reduce the above estimate.

These calculations were hopelessly optimistic about the power of a Sussex cliff to resist erosion, or that a tall cliff would somehow be better protected than a short cliff. When he wrote these words the chalk cliffs of Beachy Head were retreating 100–200 times faster than he estimated.

As he surveyed the Weald, Darwin was not looking across 300 million years of imperceptible erosion, but no more than 3 million years of rapid change, a hundred times faster.

This is not a perspective of stability but a scene of transience.

In a lengthy footnote in the second edition an abashed Darwin admitted his geological naivety:

> I confess that an able and justly severe article since published in the Saturday Review (Dec 24th 1859) shows that I have been rash. I have not sufficiently allowed for the softness of the strata underlying the chalk; the remarks made are more truly applicable to denuded areas made of hard rocks. Nor have I allowed for the denudation going on both sides of the ancient Weald Bay. . . . I can find no words too strong to express my conviction of the extreme slowness with which [cliffs] are worn away and removed.

★

> I assumed the one inch per century in order to gain some crude idea of the lapse of years, but . . . I own that I have been rash and unguarded in this calculation.

Darwin's 'rash' calculation inspired rival estimates of the duration of deep time. The physicist William Thompson (Lord Kelvin) extrapolated to the time when the volcanic Earth's surface would have been cool enough to be habitable by plants and animals and arrived at 100 million years.[4] Over the next few years, as his calculations were repeated, so the time reduced down to 24 million years.

Feeling cornered, Darwin dropped mention of his calculation of the age of the Weald, but also refused to accept the physicists' restrictions on the age of a habitable Earth. Only the discovery of radioactivity in the first years of the new century allowed geologists to recapture the initiative. With radioactive heating the thermal age calculations were almost unbounded. The Earth could be a hundred times older.

Darwin may have been wrong about the erosion of the Weald but he was eventually proved correct in his instincts about the duration of deep time.

The true erosion rates of the Weald anticline are staggering. The fastest agent of all turns out to be cliff erosion.

Over the past 150 years the white cliffs have been growing whiter, for the simple reason they have been eroding faster and plants and lichen have less time to grow. Before 1870, through 6,000 years when sea levels scarcely changed, the cliffs retreated at 5 centimetres a year. Over the past 150 years, the cliffs have been receding 20–30 centimetres each year: five times faster. The cause has been human, dredging the adjacent seabed for gravel and exposing the base of the cliffs to the direct impact of storm waves. (A team from the University of Glasgow measured this retreat from the concentrations of beryllium produced by the impacts of cosmic rays on the exposed bedrock at the base of the cliffs.[5])

Based on detailed maps dating from 1873 and 2015, the adjacent cliffs at Peacehaven and Ecclesbourne Glen have been retreating an average of 30 centimetres each year, while those at Birling Gap and Fairlight Cove have receded an annual 60 centimetres.[6]

Britain may lead the world in land loss.[7] The glacial sediment shoreline of the 40-kilometre Holderness Coast, north-east of Hull, is receding at nearly 2 metres each year: a total of more than 3 kilometres since the Romans were around. Thirty villages have already vanished.

Along 5 kilometres of coast at Covehithe, north of Southwold in Suffolk, the glacial sediment cliffs have retreated by an average of 6 metres each year for several decades: in one instance by more than 30 metres in a single winter and overall by more than 800 metres since 1800. A sea wall to protect Covehithe, it is argued, would starve the beach at the nearby resort of Southwold.

In medieval times the city of Dunwich on the Suffolk coast had a port said to rival London. Where Dingle Marshes now lie there was an estuary. Ships anchored to the lee of a N–S shingle spit known as Kingsholm. However, a series of storms in the late thirteenth and early fourteenth centuries blocked the entrance to the anchorage. Without commerce there was no money to fund seawalls. Storms devoured exposed streets and buildings until the main part of the town had to be abandoned. The Suffolk coast has retreated more than 1,500 metres since Roman times.

These instances are not going unobserved. Some 30 per cent of the English coast is suffering 'rapid erosion'. The UK Environment Agency[8] has identified 5,000 coastal properties that, without intervention, would be lost by 2030 and 28,000 by 2060.[9]

Then there are coasts where the thick clay sequences are simply sliding into the sea, as around the northern and southern coasts of the Isle of Wight, or at Lyme Regis. Along the coast to the east of Folkestone the main road and rail line run along Folkestone Warren, a chaotic landslide terrain, where Gault clay underlies chalk cliffs. On 19 December 1915 a great slide occurred along 250 metres, burying the track. A train from Ashford to Dover with more than a hundred passengers was dragged down the slope and derailed, fortunately after being stopped by soldiers waving warning lights.

We are now in the Anthropocene when human actions exceed geological processes. Long term it is sea-level rise that will cause the greatest Anthropocene impacts to Britain, where there are large areas of land close to sea level.

We will need a national memorial to commemorate the towns and villages lost to the sea. Better make sure it gets erected far inland.

At a retreat of 30 centimetres each year, in 100,000 years the chalk cliffs of Sussex will have receded 30 kilometres, inland of Lewes. In a million years, south-eastern England will have completely gone.

While the chalk cliffs are being chiselled away, there are patches of new land creation. In consuming the chalk cliffs, the sea spits out the tough flint pebbles, shaped like Jerusalem artichokes. The stones get rolled and smoothed by all the other pebbles, and bulldozed into vast mounds of tumble-rounded shingle.

Dungeness Spit is a 'cuspate foreland' comprising 500 tightly curved shingle ridges, like the growth rings on a shell. Each ridge is the product of a notable storm. The oldest dates to 3500 BCE. The spit is wearing away on the western side while growing to the east.

Old Winchelsea originally stood on a shingle spit in Rye Bay, but was obliterated by late-thirteenth-century storms, so that 300 years later cartographers did not know where it had been located.[10] At other times port towns lost access to the sea as a result of invading sand or shingle banks. The harbour at Romney had silted up by the mid-thirteenth century. The floor of the eleventh-century church is now 3 feet below street level.

Someone had the bright idea to locate two nuclear power stations on the western side of the desolate Dungeness peninsula. To sustain this location requires a never-ending 'Sisyphean' battle against the natural shingle drift. Thirty thousand cubic metres of shingle have to be dug out each year where it accumulates on the east side of the peninsula, trucked in lorries, and dumped on the western foreshore.

Other world-class shingle landforms festoon the coasts of Dorset, Norfolk, and Suffolk, the subtle product of lost chalk cliffs and violent winter storms.

Chesil Beach is a 200-metre-wide and 15-metre-high shingle barrier that runs for 29 kilometres in a SW-facing concave curve along the Dorset coast, from West Bay at Bridport to the isle of Portland. Ridges on the seaward side mark the latest storm tides, with the record wiped clean when a higher storm tide arrives. On a November night in 1824 the sea burst through the shingle barrier close to the south-east end, destroying eighty houses and taking up to fifty lives. The flint and chert quartz pebbles are winnowed by the waves, from pea-size at West Bay up to tangerine diameter at Portland, such that (it was said) a local fisherman arriving on the beach at night could locate their landfall from the musical pitch of the gravel.

To find a landscape-art rival to the Zen gardens of Kyoto, with their gravel raked to resemble waves among rocky islands, there are the wave-sculpted raised beaches along the remote west coast of Jura island in the southern Hebrides (Fig. 70). The storms are more ferocious than on the coast of Dorset, and the winnowing of quartzite boulders extends up to pebbles the size of curling stones. With the melting of the Scottish ice sheet the land has been rising, so that the

Figure 70. Raised quartzite cobble beach, west coast of Jura

beach ridges reach much higher than even the greatest modern storm. The quartzite is free of any other minerals, vital for plant life, leaving the shingle ridges bare and sterile, except where adorned with giant vegetation mat rosettes (nucleated from the nutrients brought by a carcase?).

★

The hardest thing of all to preserve is a landscape. Like a half-licked ice cream, the landscape continues to be eroded away and eventually will have gone. Preservation requires some extremely rapid change, the equivalent of freeze-drying.

West of the Shetlands, during the Paleocene, first the sea floor rose by an estimated 550 metres to create new land. A branching pattern of streams and rivers developed. Then within 2 or 3 million years the land sank down again[11] and this pattern of river valleys became preserved, buried beneath layers of marine sediment.

Around Britain the land already lost had landscapes as complicated and varied as the land that remains. When all we find are the vestiges, it is left to our imagination to recover what has already gone. On the north side of Tory Island off the north-west coast of Ireland there are 85-metre-high cliffs beyond which the land is consistently tilted down to the south-west.[12] Stand at Ballynagall on the east shore of Smerwick harbour in County Kerry and you will see the remains of a SW-flowing river valley for which all the former SE-sloping tributary valleys are beheaded by sea cliffs.

Two detached incised meanders survive from a lost river valley heading parallel to the coast, south of Hartland Quay in north-west Devon.

On the south-west corner of the Hebridean island of Eigg, there is a tantalizing fragment of a landscape tableau formed towards the end of the volcanic activity 59 million years ago. A steep river channel had cut its way down through the pile of lavas, flowing towards the NNW. The channel was floored with large boulders, including those sourced from ancient basement rocks exposed on the far side of the

Camasunary Fault. An eruption of glassy 'pitchstone' (granite composition) magma has filled the valley, cemented the boulders, and preserved the underlying landscape. The 'landscape preserver' has now become the landscape, forming Eigg's summit: the 'Nose', reaching 393 metres above sea level.

★

We have reached the end of the story, not because we have come to a natural terminus but simply because this is where we are today. Tectonic history has run out, but this will not be the end of the tectonic story.

Let's bring Charles Darwin back to have a near-final word. After viewing the recently destroyed city of Concepción in Chile, in 1835 our accidental geologist wrote:[13]

> The earthquake alone is sufficient to destroy the prosperity of any country; if beneath England a volcanic focus should reassume its powers; how completely the whole country would be altered. What would become of the lofty houses, thickly packed cities, great manufactories, the beautiful private and public buildings? If such a volcanic focus should announce its presence by a great earthquake, what a horrible destruction there would be of human life. England would become bankrupt; all papers, accounts, records, as here, would be lost: and government could not collect the taxes... and failing to maintain its authority, the hand of violence and rapine would remain uncontrolled. In every large town famine would go forth, pestilence and death following in its train.... Who can say, how soon such will happen?

Darwin was gloomy. Perhaps such a disaster could occur in Britain? Geology in the 1830s seemed open to the possibility. (In subsequent editions of his *Voyage of the Beagle*, having become less confident about the 'volcanic focus', he substituted 'subterranean forces'.)

On 2 October 1836 the twenty-seven-year-old Darwin returned from his five-year circumnavigation high on his wild and illuminating adventures. In July 1838, he sought to recreate some of that 'Beagle' spirit through mounting a lone geological expedition to the Scottish Highlands to visit the famous Parallel Roads of Glen Roy. After the camaraderie of ship-life, he was lonely. In the same month he drew up a list of the pros and cons of marriage. Four months later he was engaged and within another two months married—this was to be his first (and last) solo 'big adventure'.

The railway journey from London to Liverpool, with its two changes of line, had only opened a year before. From Liverpool he took a steamer to Glasgow pleased, as all around him succumbed, that his 'Beagle sea-legs' had finally immunized him against seasickness. From Glasgow to Edinburgh and on to the Highlands he travelled by gig and cart. The five days he spent around Glen Roy were in beautiful and uncharacteristic sunshine.

His memory returned to the eternal sunshine of central Chile and the experience of another staircase of raised beaches at Coquimbo, where five or six shingle-covered terraces extended up the valley for 60 kilometres from the modern coast, rising to a height of 75 metres, each covered with shells still found on the beach below. The beaches reflected long periods of stability between episodes of the emergence of the continent, which, as he had seen at Concepción, accompanied great earthquakes.

In Glen Roy, Darwin studied the three horizontal rims around the valley, once mapped as trackways, or deer-hunting courses for ancient Scottish kings. He dug into the material and found it to be composed of the pebbles and sands of a beach. Other geologists before him had considered these shorelines were from some lake whose natural earth embankment had subsequently disappeared. For Darwin, from Coquimbo to Glen Roy, what other explanation could there be than that these recorded past sea levels? At elevations of 260, 325, and 350 metres, these manifested far more uplift (or 'emergence') even than Coquimbo. If the raised beaches were like those of Chile, could that mean a similar role for tectonics and great earthquakes, or was this like the silent emergence of Sweden?

Only one detail nagged at his interpretation of Glen Roy's roads: why were there no accompanying seashells? As he wrote to Lyell a week after returning: 'I have fully convinced myself (after some doubting at first) that the shelves are sea beaches—although I could not find the trace of a shell.' Darwin wrote up his interpretation for publication in the eminent *Philosophical Transactions of the Royal Society*.

Four years later in a letter to his second cousin William Darwin Fox, clergyman and naturalist, he acknowledged:

> My marine theory for these roads was for a time knocked on the head by Agassiz ice-work—but it is now reviving again—I don't mean, that I ever doubted, but others did (even Lyell for a time became a catastrophist)—and they have now gone back to the elevation theory.

Agassiz had suggested that the lakes had been dammed by an ice sheet. Darwin's interpretation of the Parallel Roads of Glen Roy as the result of recent land emergence 'confirmed' an idea that Lyell had first proposed in the third volume of his *Principles of Geology*—that erratic (far-travelled) boulders found along the coast, or in the mountains, had been carried by icebergs spawned from mountain glaciers. For icebergs to perform this function would require deep water. Over the next twenty years, in Geological Society discussions, Lyell's land emergence and iceberg explanation provided a successful antidote to Agassiz's 'catastrophist' Ice Age theory.[14]

Darwin found additional support for the idea that the mountainous regions of Britain had only recently emerged out of the sea from reports of deposits containing modern shells, more than 400 metres above sea level on Tryfan mountain in north Wales.[15] He visited the location in 1841 (but could not find any shells).[16] (We now understand that an ice sheet had scraped shelly sands from the edge of the Irish Sea and pushed them high up on land.)

In 1861 the Scottish geologist Thomas Jamieson, advised by Darwin and Lyell, made two visits to Glen Roy. He returned with a detailed explanation for how the shorelines formed from three levels of a transient lake dammed by a downstream glacier. Darwin, always ready to self-deprecate, acknowledged privately that his own paper was 'one long gigantic blunder'.[17] So ended the idea that Caledonia was 'Chile'. Yet twelve years later[18] (in the latest edition of his book *The Antiquity of Man*), Lyell still sustained some wild 'tectonic' notions as an alternative to an Ice Age: that Scotland had recently been submerged by 2,000 feet, along with England and Wales by 1,300 feet, 'north of a line from London and Gloucester', while the land south of the Thames 'alone remained above water'.

Over the following century the opposite idea was to become the orthodoxy. The idea that Britain was a bedrock of stability.

★

Henri Baulig, born in Paris in 1877, graduated from the Sorbonne and, in 1904, received a French government scholarship to attend Harvard University. There he studied under William Morris Davis, who had set out the founding principles of the geomorphology of river systems. In 1928, back in France, Baulig finally completed his PhD, 'On the Origin of the Landforms between the Central Plateau of France and the Mediterranean'. Receding global sea levels, he believed, had left their mark in raised beaches and abandoned cliff-lines. Based on the prominent raised beaches found around the coasts of Calabria and Sicily he named the associated 'sea levels': 'Calabrian' at the highest level above 200 metres, 'Sicilian' at 100 metres, 'Milazzian' (after a town near Messina) at 60 metres, 'Tyrrhenian' at 30 metres, and 'Monastirian' at 18 metres (from a town on the north-east coast of Tunisia). Baulig believed that the sea levels had consistently fallen since the Pliocene, probably as a result of the unseen lowering of the ocean floor. From the height of some trace of a former sea level, you would know its age.

The idea was imported to Britain through the geologist Sidney Wooldridge, born in 1900. In 1928 Wooldridge identified a former 200-foot 'Milazzian' sea level around the London Basin while also claiming a pebble gravel surface found in south Hertfordshire at 370–500 feet was the 'Sicilian' sea level. An older marine platform at 550–650 feet was potentially 'Calabrian'.

Through the mid-twentieth century the idea became mainstream: that Britain was a bedrock of stability into which receding sea levels had carved a staircase of rock-cut platforms and contemporary beaches.

Wooldridge became Professor of Geography at King's College, London. British geomorphologists discovered more and more examples of summit plateaus and topographic benches, from Cornwall to Cumbria. Miocene age marine rock-cut platforms have been 'surmised'[19] in the 600-square-kilometre 'Menaian' Surface of Anglesey, in the 'Bosherton-Castlemartin Plain' of southern Pembrokeshire, in the 'Reskajeage' Surface of west Cornwall, and in the 'Durdham Down Surface' around Bristol. The Channel Islands are planated, as is the coastal peneplain of southern Ireland. In east Cork and Waterford there is a staircase of planation surfaces.[20]

In 1972 the French geographer Annie Reffay poked fun at the British obsession with this 'jeux des palliers' (the game of levels), which was perhaps unfair as the game had been introduced to England by a Frenchman.

(In contrast, the French have their own obsession with neotectonics—convinced, often from meagre evidence, that some fault such as the Pays de Bray axis is 'active'.[21] The southern 'continuation' of the Sticklepath Fault across the Channel into the Quessoy–Vallet fault zone has been claimed, for example, to control the 'ongoing uplift' of western Brittany.[22])

It was not until the 1960s that the whole 'game of levels' collapsed. The raised beaches of Calabria and Sicily were discovered to be situated on coastlines with the most rapid rates of tectonic uplift (at 100 metres for each 100,000 years) in the whole Mediterranean region. These beach elevations revealed nothing about global sea levels.

There has never been a sea level 200 metres, or even 100 metres, higher than today, at least for the past 60 million years. The landmass of Britain has been far from stable. We cannot assume any tableland is at the elevation it was when it was carved by the waves. Other former wave-cut platforms have been left tilted.

Yet the mid-twentieth-century idea of immovable Britain, at its peak around the time of the Second World War, now seems something of a metaphor. The high tide in this theory of receding sea levels and rock-steady Britain was also the peak of empire.

At the same time, most geologists were resistant to the idea of continental drift. The bedrock land had moved neither horizontally nor vertically. Through the 1960s and 1970s, as geophysicists called the shots over plate tectonics the Geological Society of London left the job of organizing a meeting on the new earth sciences paradigm to the Royal Society.[23]

We now realize that throughout the past 66 million years, this land has been far from quiescent. It has been split by magma-filled cracks, wracked by great earthquakes, and spindled by faults, short-circuiting between the Alpine and Atlantic plate boundaries. Perhaps we need to develop a more realistic appreciation of unstable Britain, tectonically buffeted and reshaped, since this region first emerged 60 million years ago.

★

Does geological history ratify Britain's island status or highlight that this is really just a north-western peninsula of Europe?

Since its emergence some 60 million years ago, Britain's insular status has alternated. An island through the first half of the Eocene and the Pliocene. Connected to Europe through the later Eocene, Oligocene, and Miocene. Over the past 60 million years Britain has perhaps spent twice as long connected to as independent from Europe. There is a metaphor for both isolationist and European persuasions.

If in a 'deep-time counterfactual' the geological history was rerun without the most aggressive phase of the 450,000-year-old Anglian glaciation, today you could probably still walk along the North Downs all the way to France. The rising North Sea lake must have come close to floating the ice sheet to the north so that all the trapped water flowed out to the Atlantic. And then there would be no Dover Straits. No need for a Channel Tunnel or a port at Dover. With today's technology there would be motorway and rail links across the isthmus and a history of massive fortification to try (unsuccessfully?) to prevent invaders. A sea-level shipping canal, already twice dug deeper and wider, cuts through the isthmus (in a deep cutting like the Corinth

Canal) so that ships destined for ports around the southern North Sea can avoid the long Scotland circuit.

This link would, however, be geologically precarious. Storms arriving up the Channel would devour the weak Jurassic and Early Cretaceous sediments, before nibbling away at the chalk cliffs on the south side of the isthmus. The north side would be exposed to winter northerlies. Sooner or later, the isthmus would fray to the thinnest wall of chalk and then the sea would cut through. Unless heavily reinforced along the shoreline, Britain's resumed island status would eventually be inevitable.

Was the separation of the island of Ireland from the rest of Britain also inevitable? Between Wales and Ireland any Miocene topography in the weak Oligocene sediments is long eroded away. Farther north it is only the deep gouging from the ice sheets following older river valleys that has separated Kintyre and the Mull of Galloway from Antrim. Sixty million years ago the volcanoes saw no great distinction between Scotland and Ireland.

As for Scotland and England, they have shown no inkling of partition through the Cenozoic. One has to go back more than 425 million years to find the two kingdoms on different plates, separated by a wide ocean.

In November 2023 I asked the large language AI model ChatGPT to summarize the tectonic history of Britain through the Cenozoic and learnt in response that Britain had been drifting away from Europe since the Eocene, the separation accelerating since 2000 (as drift of continents elided with drift of sentiment).

★

In unravelling this story, we have identified nine tectonic episodes in the prehistory of Britain and Ireland since the death of the dinosaurs. In order there were:

- 'The big tilt'—the dome uplift of the British Isles through to the first volcano;

- 'The starter ocean'—the configuration of spreading ridge dykes and lavas from Skye through to north-west Ireland;
- 'Giant dykes'—the long-distance dykes that set out to split Britain in two;
- The initial opening of the Norwegian Sea that brought sarsens and London clay;
- The Pyrenean orogeny comes to southern England and south of Ireland;
- The opening of the Rhine Graben;
- The Celtic rifted basins, when rifting linked from the Mediterranean to the Norwegian Sea;
- 'Caught in the jaws of a tectonic vice'—the compressional tectonics of the Miocene;
- The Pleistocene tectonic see-saw between rising England and the sinking North Sea.

To which list of nine I need to add a tenth tectonic episode at the beginning of the Paleocene.[24]

The plume volcanism that kicked off this story began at 62 Ma. Volcanic fires broke out simultaneously on both coasts of Greenland as well as through western Britain. Four million years had passed since the asteroid collision that initiated the Cenozoic. For 10 million years after 62 Ma the African plate ceased colliding with Europe (until contact was resumed again along the Pyrenean mountain range). Before 62 Ma, at the start of the Cenozoic, the collision with the African plate was compressing the crust across western Europe. Around the southern North Sea and passing into central England, a series of sedimentary basins, probably including the Weald Basin, became squeezed and rose up by many hundreds of metres.[25] It was probably this uplift that erased the record of the atomized asteroid. At 62 Ma the forces of compression were relieved and the tectonics switched into mild extension and dyke swarms.

Four lines of NW–SE faults run across Carlingford and Slieve Gullion volcanoes. Unlikely to have all been active in concert, there were even more tectonic subchapters than in my headline index.

The theory of plate tectonics, developed at the end of the 1960s, explained how continents drift, how new ocean crust forms at spreading ridges and eventually returns back to the mantle down subduction zones. Plates are broad and rigid, which is why plate boundaries maintain a strict geometry as they channel their movement.

Over the past 66 million years, on several occasions nascent, or 'tributary', plate boundaries have passed through Britain.

The 'starter ocean' was an attempt to form another 'spreading ridge', through north-west Britain and Ireland. Then for a period, the intrusion of the giant dykes attempted to open a new spreading ridge that would split Britain in two. Around 45 Ma there was the 'North Pyrenean' band of tectonics, which spanned from the Weald, through the North Celtic Basin and west into the North Atlantic, a spillover from the main plate boundary collision zone through the Pyrenees and along the north Biscay coast of Spain. Bounded to the east by the Rhine Graben rift, France was for a period its own tectonic plate.

And last, there was the late Oligocene (25 Ma) garland of Celtic basins through western Britain and northern Ireland, part of a chain of rifts and strike-slip faults that linked together from the Mediterranean to the Norwegian Sea.

Although each tectonic episode only achieved a few kilometres of movement (and the giant dykes opened by no more than 200 metres), in terms of their continuity, and their consistent geometry, each played by the rules of plate boundaries.

Why does tectonics follow tortuous paths through the continents? Because faults navigate around stronger areas of continental crust, known as cratons. Cratons are tough and gnarly, deeply rooted into the underlying colder mantle, exposing crystalline rocks originally formed tens of kilometres underground.

As noted in Chapter 6, beneath central England lies the London–Brabant (or Midlands) Platform. The Weald Basin to the south was diverted round this craton. Even the pattern of modern earthquakes across the Welsh borders seems to be following the NE-trending cratonic boundary.

The Sticklepath Fault has proved to be a versatile, much-travelled, tectonic highway, active in three different periods of our story. This is because it skims past the north-west edge of the great 'cratonic' granite which underlies the whole of the south-western peninsula from central Devon to the Scilly Isles. Almost all the passing tectonic 'traffic' has been guided along the Sticklepath Fault.

The Sticklepath Fault runs for 300 kilometres from mid-Channel to near Carnsore Point, the south-east corner of Ireland. Carrying several kilometres of displacement, it must plug into a plate boundary—but where?

Through the Cenozoic the fault has moved in lockstep with the NNW-trending Codling Fault. Wherever we can measure the Cenozoic dextral displacement—in the middle of Devon, offshore Dublin, or through the Carlingford Volcano—it is much the same: 4 or 5 kilometres (a 'Park Run'): maintaining consistency, just like a plate boundary, or the 'crack in the wall'.

From dating the Sticklepath Fault 'pull-aparts' we even know when the fault was moving: from 52 to 25 Ma.

Sticklepath and Codling are parts of the same fault system. The Codling Fault runs onshore through the Carlingford Volcano and can be chased on-land along the Portrush Fault, altogether adding another 300 kilometres.

Impressed by its continuity we made a bold extrapolation (see Fig. 37). The fault sets a course a few degrees west of north through Blackstones and on to St Kilda Volcano. The remotely mapped sea-floor geology does not recognize such a fault, but horizontal displacement is easily missed. Like the crack in the wall, this fault must continue. Project this line another 250 kilometres to 60°N latitude and we arrive at a contemporary plate boundary that could finally explain the very existence of our Sticklepath–Codling–Portrush Fault.

This plate boundary was the former NW-trending West of Shetland transform fault. In a story already told, around 50 Ma, spreading in the Norwegian Sea transitioned to east–west, leaving the original west of Shetland transform fault in an impasse of oblique collision, piling up two parallel tortured underwater mountain ridges, named Wyville Thomson (after a nineteenth-century Scottish marine biologist) and Ymir (after a primal being of Norse mythology).[26] Our projected fault intersects the south-east end of the Ymir Ridge. And there, in this cauldron of stresses, I believe, lies the explanation for the fault's creation, cracking through the plate for 1,000 kilometres. It took until 25 Ma before the blocked transform was finally bypassed.

As to how the southern continuation of the Sticklepath Fault traversed France and connected with an Alpine or Pyrenean plate boundary, I leave that to a reader to solve: a puzzle in tectonic 'train tracks'.

Meanwhile the 'world-class' Sticklepath–Codling–Portrush fault system is known (Lundy, Carlingford, Slieve Gullion) or surmised (Blackstones and St Kilda) to have intersected all the extinct western ley-line volcanoes, each a thermal 'weakest link' crustal perforation.

Even in configuring its tectonics, this was a volcanic isle.

★

When I started writing this book, I had no idea that Charles Robert Darwin would make even a single appearance, but he kept turning up at key moments in the narrative. This leads me to want to write a report card on him, not as the justly celebrated, world-famous author of a theory of evolution by natural selection, but in the role he first played: as a geologist.

Darwin was overflowing with ideas and interpretations, but by no means was he always correct. He exhibited great field skills in finding the Butterton Dyke. He also developed an original and correct perspective on coral reef islands and a sinking ocean floor. However, his interpretation of the parallel roads of Glen Roy and his age estimate for the denudation of the Weald were quite unrealistic. Yet he was always quick to admit the error of his ways when confronted with a superior theory. In his autobiography he acknowledges:

> I have steadily endeavoured to keep my mind free so as to give up any hypothesis, however much beloved (and I cannot resist forming one on every subject) as soon as facts are shown to be opposed to it. I have had no choice but to act in this manner, for with the exception of the Coral Reefs, I cannot remember a single first formed hypothesis which had not after a time to be given up or greatly modified.

In the spirit of Darwin I have made a list of loose ends and unexplained phenomena, identified in the course of researching this book, from which I have selected ten tectonic puzzles waiting to be solved. Here goes.

(1) Where did the Antrim dyke swarms and lava flows come from? Just as in Iceland, on Skye and Mull we can show the dykes and lavas were supplied by a big central volcano, situated in the middle of one edge of the lava field. But there is no central volcano in a location to feed the dykes and lavas on Antrim. There must instead be one or more buried magma chambers. As the chamber emptied to supply a dyke, the overlying rocks must have fractured and subsided. There has to be evidence of such caldera collapse waiting to be mapped.

(2) On the north-west coast of the island of Jura the basaltic dykes, fed from Blackstones Volcano, stand up like great sculptures and walls, even though the surrounding rock is itself resistant to erosion. Yet in most dykes the minerals rot and the course of the dyke becomes a ditch. What is special about these dykes which makes them so resilient?

(3) The mountain ranges of western and northern Norway reach heights of more than 2,400 metres. On the east coast of Greenland they reach more than 3,000 metres. And yet while the bedrock is ancient these mountains are geologically young. They seem to have been raised passively, without crustal tectonics. The only credible explanation is that a vast 'pillow' of lower-density

material has arrived beneath them. What is this material? If it is magma, why didn't it break through to erupt at the surface? If it was originally extra-hot, why has it not cooled, contracted, and the uplift reversed?

(4) What determined the two 'ley lines' of volcanoes across Britain? The rocks of the mantle are supposed to be too hot and malleable to sustain a crack, for the same reason you can't sustain a crack through honey. What else could these boundaries be, up which, at regular intervals, magma has emerged? Maybe this marks the path of a deep 'super-dyke', from which the volcanic centres were sourced? It seems too straight for a single fault cutting through all the geological structure. The orientation of the ley lines is picked up by other tectonic elements: in the broad underground dyke north of Skye, in the Portrush Fault, and in Darwin's Butterton Dyke.

(5) We can establish that the giant dykes in north-east England originated at the Mull Volcano, whereas those to the south-west in Anglesey and north Wales came out of Slieve Gullion Volcano. However, there is one giant dyke in the middle, first encountered on the Isle of Man, as well as many standard-sized dykes. And then there is the different geochemistry and orientation of the Butterton Dyke. Where did they all originate?

(6) What has caused the southern North Sea to founder over the past 2.5 million years? To sink like this implies some additional load beneath, a downward current in the mantle, or a mineral phase change as the mantle cools. Will it continue?

(7) Why are there no sediments in Britain that record the boundary between the Cretaceous and the Cenozoic? Was no sediment deposited because it was already all land, or did a tsunami sweep over the emerging landmass and its fringing shallows?

(8) Many of the principal faults of northern England around the Pennines look geologically young, as though reactivated late in their history. How do they fit into the tectonic story?

(9) Why were there great earthquakes in northern Sweden and Norway 9,000 years ago, just as the last remnants of the ice sheet melted? Nothing of this character has been found in other regions which formerly lay below thick ice sheets.

(10) How in 1755 did a great rupture offshore from south-west Portugal generate an earthquake as large as magnitude 9? All other earthquakes of this size over the past 400 years have been located on subduction zone plate boundaries, mostly around the Pacific. Where was the 1761 earthquake situated and how does it relate to the great 1755 shock? Is there a potential for such an earthquake farther north, closer to Ireland?

13

The volcanoes return?

Britain has seen many volcanoes in the geological past and the region will likely see more volcanoes in its future. Seven generations of dyke swarms have been identified in Scotland. Northern Britain was peppered with volcanoes between 400 and 300 Ma, leaving natural fortress landscapes, as at Castle Rock, Edinburgh and Bass Rock in the Firth of Forth. The great 300 Ma basaltic Whin Sill creates cliffed topography followed by Hadrian's Wall, High Force waterfall on the River Tees, three magnificent coastal castles, and the Farne Islands.

Whereas there was little volcanic activity in the 200 million years before the Paleocene revival, in the mid-Jurassic (166 Ma) a band of WSW–ENE fissure eruptions and small volcanoes, fed by dykes, extended for 25 kilometres through the middle of the North Sea.[1]

All that remains of an isolated Early Cretaceous (131 Ma) volcano, located 15 kilometres south-west of Land's End, is a rocky reef protruding a few metres out of the waves, a dangerous obstacle in a busy shipping lane. Named 'Wolf Rock' on account of the wind's howling, the reef merited a warning sign. In the 1790s a metal sculpture in the shape of a wolf was fixed on a pole. After a succession of equally short-lived structures were removed by winter storms, in the 1860s a strong 36-metre-high granite lighthouse was constructed. Now automated, solar-powered, and fitted with a helicopter landing pad, that lighthouse survives today.

Wolf Volcano is at least 10 million years too old to be the origin of the montmorillonite clay layers, derived from volcanic ash, found in Lower Cretaceous strata. Mined for hundreds of years in Bedfordshire, a 1793 advertisement reported:[2]

> FULLERS EARTH—ALMOST equal in Quality to Soap, particularly for scouring Coarse Cloths, Blankets, &c may be had from the finest Vein in the World, at 6d. per Bushel, or Hundred Weight, at the Pit at Aspley, near Woburn, Bedfordshire.

Also mined at Redhill, Surrey, 'Fuller's Earth' was originally volcanic ash deposited from nearby eruptions. However, the location of the volcano source has never been found.

Rerun with today's cities and industry, Europe's most costly geocatastrophe since the end of the last Ice Age was a volcanic eruption only 500 kilometres from London.

The Laacher See eruption of 13,006 years BP[3] blew off the top of a mountain in the Eifel region, Germany, 24 kilometres north-west of Koblenz and 37 kilometres south of Bonn. Laacher See is the lake that filled the 2-kilometre-diameter caldera, named after an eleventh-century Benedictine abbey founded on the shoreline. (Volcanic calderas, testament to a violent past, seem to attract religious contemplation; the pope has his summer residence on the shores of volcanic Lake Albano, 3.5 kilometres across, situated 25 kilometres outside Rome.)

The Eifel volcanoes first surfaced 700,000 years ago. The West Eifel volcanic field contains 240 cones and flooded crater lakes whereas the

East Eifel field contains about a hundred eruptive centres, of which Laacher See is among the youngest and most violent.

Two hundred thousand years ago the Bellerberg Volcano, 10 kilometres south of Laacher See, erupted three streams of magma, one running for 3 kilometres into the valley of the Nette at Mayen. The solidified lava was a 'phonolite', low in silica and 'mafic' (dark, iron and magnesium silicate) minerals, but rich in nepheline and potassium feldspars, much the same composition as Vesuvius magma (and found uniquely in Britain at Wolf Rock).

The name phonolite (meaning 'sound-stone') is on account of the rock's metallic musical tone. Walking on phonolite scree tinkles like a gamelan, perfect for the stone xylophone—the 'lithophone'. The German name is the onomatopoeic 'klingstein'. Geologically so young the rock had not started to decompose, Bellerberg phonolite was the perfect medium for manufacturing grinding stones. Easily quarried because the lava flows had developed strong columnar jointing, the mining of millstones from Bellerberg quarries pre-dated the Romans. Close to the Rhine the blocks were transported by river as far as France and Austria and eventually by sea to England. This was the source of the lava that drove the Roman manufacturers of Hertfordshire puddingstone querns out of business.[4]

The 13,077-year-old Laacher See eruption occurred in late spring or early summer. We know because of young fruits and foliage, even a foal's hoofprints, preserved beneath the ash.[5] Like the 79 CE eruption of Vesuvius that destroyed Pompeii, not only were there thick falls of ash, but when the eruptive column was most intense and high, it collapsed to generate pyroclastic flows, red-hot avalanches of still-molten ash, flowing over the landscape, fusing into thick drifts of 'ignimbrite' rock (from the Latin *ignis* 'fire' and *imber* 'storm cloud'), and burning everything in their path. However, whereas the 79 CE eruption emitted a cubic kilometre of ash, the Laacher See eruption produced 20 cubic kilometres.

In terms of eruption volume and materials, as well as the size of the resulting caldera, Laacher See has much in common with the largest on-land eruption of the last fifty years, in 1991 at Mt Pinatubo in Luzon, Philippines. The two eruptions even occurred at the same time of year.

At Pinatubo the earliest portents were steam eruptions accompanied by tremors in early April. The main phase of cataclysmic eruption took place over four days in June, and involved a series of explosive eruptions, the highest reaching 34 kilometres altitude, laden with so much ash that sections of the column collapsed to generate pyroclastic surge avalanches that flowed for long distances over the landscape. At both volcanoes the final explosions blew off the mountaintop and left a caldera. For years, all around Pinatubo rain turned thick dunes of ash into lahar mudflows.

Around Laacher See, more than 8 metres of ash accumulated within 5 kilometres of the vent, dunes reaching 50 metres thick close to the crater. Trees, laden with ash, were felled by explosive blasts and downdrafts, out to 4 kilometres. Through the main phase of the eruption, it seems a low-pressure system passed by; the winds blew strongest and longest from the south-west but then veered round to the north behind the front. As a result, the 1-metre contour for ash-fall extends 40 kilometres to the south but 120 kilometres to the ENE and only 5 kilometres to the west.

The first of the pyroclastic flow avalanches at Laacher travelled north-east down the valley of the Brohl River to where it joined the Rhine, 6 kilometres from the volcanic vent. The avalanche deposits were 60 metres thick in the Brohl valley, and travelled to the far side of the Rhine, leaving a broad dam of fused ash at least 20 metres high. Masses of tree-trunks and ash-fall mudflows accumulated behind the dam for 7 kilometres, to the Andernacher Pforte gorge.

Layers of pumice reveal the resulting lake reached up to 15 metres higher than today's water levels, backing up the Mosel and extending over at least 80 square kilometres.[6] With raised flow from the spring thaw in the Alps the lake probably took five to seven days to fill.

Downstream of the dam, the bed of the river dried up completely between the Brohl and Ahr side streams. Under today's conditions, for all its length to the sea any boats and barges would have been stranded on the mudflats. Any industry or power station dependent on cooling water from the river would be in trouble. (Fortunately, the nuclear power plants once located along this section of the Rhine are now closed.)

Once the water overtopped the obstruction, the dam seems to have failed rapidly, allowing 1.5 billion cubic metres of water to surge down the river. Today, such a torrent would flood riverside towns like Bad Breisig and Remagen, the riparian destruction passing all the way to Bonn.

In comparison with Pinatubo it seems the explosive phase of eruption lasted longer than four days. Thirty episodes of pyroclastic flows have been identified and the sediments deposited by flooding when the dam collapsed are covered in ash-fall, suggesting that the main phase of the eruption lasted at least seven days.

How to estimate what would be the costs of a repeat of the Laacher See eruption today? First, there is the direct building damage from the ash-fall. Four metres of ash would have fallen on the neighbouring towns and 150 centimetres accumulated on Koblenz (population 114,000). More than a metre of ash is enough to collapse most roofs, especially once loaded with rain. For the 1,300-square-kilometre area including many towns and villages that experienced more than a metre of ash, one might expect many total losses—roads impassable, constant mudflows, drains blocked, forestry ruined, livestock killed, fields and foliage shrouded in ash. Unliveable conditions for many years. And then there would be the damage to buildings and equipment for the 300,000-square-kilometre area with lesser amounts of ash-fall, from central France to northern Italy, western Poland, and southern Sweden. And that is before we include all those thousands of properties that would have been inundated by the Rhine Lake or washed away in the great Rhine flood that followed the collapse of the barrier.

Added to which would be the disruption of all river transport; the evacuation of more than a million people within 40 kilometres of the volcano for several weeks, and more who lived along the River Rhine in Bonn or Koln; the disruption to all farming and forestry; the blocked drains and downed power lines; the lahar mudflows that continued for years. The total bill would be well in excess of 200 billion euros.

Beneath the clouds of ash carried to the north-east and south, the eruption would have brought periods of total darkness through the early summer days. This awesome catastrophe is known to have had a dramatic impact on the tribes in the region.[7]

The average time separation between Eifel eruptions has been around 2,000 years. GPS data show the surrounding land is rising by 1 millimetre per year and also extending: a dome above an underlying mantle plume, unique across north-west Europe away from the influence of the former ice sheets.[8] The longer the wait—already four times the average interval—the bigger we can expect the next eruption to be. Volcanic gases bubble through the Laacher lake while deep earthquakes reveal the movement of underlying magma.

Could such a volcano surface in England? Britain is the only major country in western Europe currently without active volcanoes: France, Germany, Italy, and Spain all shelter volcanoes on their mainland territory, which have erupted since the start of the Neolithic (12,000 years ago) and may have figured in Stone Age mythologies about angry gods and the underworld. Besides the usual plate tectonic reasons for volcanic activity, either at spreading ridges, as in Iceland, or above subduction zones, as through the Greek islands and along the west coast of central Italy, there are active volcanoes in the eastern foothills of the Spanish Pyrenees and around the Massif Central in France. At the northern end of a 300-kilometre-long volcanic zone, which reaches almost to the Mediterranean, the youngest Chaîne des Puys (Monts Dômes) includes a hundred cones, domes, and maars, stretched on a 40-kilometre N–S zone, no more than 5 kilometres wide, passing to the west of Clermont-Ferrand. The most recent eruption occurred 5,840 years ago, surely witnessed by our Neolithic ancestors.

At the northern end of the Rhine Graben, volcanoes were active through the Miocene, before a gap of a few million years and then the arrival of the Eifel craters.

The only feature these locations share is proximity to Oligocene rifts. These rifts also continue through the Celtic basins of western Britain, so one might ask: why were there no accompanying volcanoes?

And given a principal indicator of activity in the Eifel volcanic province is ongoing uplift, why is the rise of central southern England not a potential indicator of future volcanic activity? We find the incised meanders of the Mosel next to the Eifel volcanoes, but what of the incised meanders of the lower River Wye?

The youngest British igneous activity found (so far) was 45 million years old, 10 million years after the last eruption in the Hebrides: a single mantle-derived dyke, on the western edge of the island of Lewis.[9] Such was the research fascination with this petrological candy store of gem-quality sapphires, and a diverse range of mantle 'xenoliths', that the outcrop is now exhausted.

In 1897, in the lead-up to the turn of the century, with a taste for the apocalyptic, the pages of *Strand Magazine* recounted a dramatic and deadly outpouring of basalt magma, issuing from a 13-kilometre-long fissure through Marlow, Buckinghamshire, comparable to the 1783 Laki, Iceland eruption, or the Snake River basalt flows of Idaho. Over a few days lavas filled up the Thames Valley, burying Henley and Maidenhead, spreading east over the lower parts of London, as well as damming two great lakes upstream: Lake Oxford and Lake Newbury. The hero, who first viewed the Marlow eruption, managed to escape by pedal bike, keeping to higher ground, in time to warn his wife and children in Bayswater to escape to the uplands of Hampstead. With Westminster buried, parliament relocated to Manchester.

Such a flood basalt eruption is scientifically implausible, so far from a plate boundary or away from a mantle plume. An isolated volcano would have a composition and eruptive repertoire like the volcanoes in Germany.

If a new volcano appeared in southern Britain, scientists would soon discover evidence consistent with this outcome.[10] Seismic data already reveal a 'low velocity zone' (indicative of melting) in the mantle 100 kilometres beneath western Britain, as exists beneath other West European volcanic provinces.[11] How might it begin?

It was January 2032 when the residents in Ascott-under-Wychwood in north Oxfordshire first felt a series of tremors. By late spring, the tremors had intensified and were being felt in Charlbury and Chipping Norton. The British Geological Survey had installed a network of seismic recorders around the villages and discovered the shocks were being triggered in the middle of the crust, but that they were growing shallower. Tremors were also being recorded between Reading and Newbury, at the site of a pronounced geomagnetic anomaly, known to be from a deeply buried igneous intrusion, perhaps geologically much younger than had been assumed.

Then, on a frosty morning towards the end of 2033, a farmer in Chadlington noticed steam rising from a spring in a corner of one of his sheep fields. He went to inspect and found the water was warm. Two weeks later he returned and found the water was hot, staining the grass brown.

The shallowest tremors were now less than a kilometre underground. More warm springs were found north of Charlbury, several with bubbles of gas smelling of rotten eggs. It was on a summer night that a passing driver noticed something extraordinary—a rift had opened in a coppice, from which smoke was billowing. The smoke was hot and filled with cinders. Several trees had caught fire. The police first set up a 200-metre-diameter cordon, as more and more ash poured from the chasm. News crews from across the globe had come to film what was under way. This was the first new volcano in Europe since 1538.

The eruption was growing in size and power. Within the first few days a conical ash hill had grown around the eruption vent, rising 20 or 30 metres. After a week the police expanded the cordon to more than a kilometre, including in the evacuation zone six farms and a couple of small villages. Two months later the cone is 90 metres high and spans 2 kilometres across at its base, while the evacuation zone extends for 5 kilometres all round the volcano. People have been told to keep their windows closed in Oxford and Woodstock. Rising magma has set fire to underground coal seams. In the absence of oxygen the smoke is rich in toxic carbon monoxide. There are questions whether to drill down to try to stop the combustion spreading. The town of Chipping Norton will have to be evacuated. No one can give an answer as to when the residents will return.

Unlikely maybe, but not implausible. What will be the next tectonic episode to affect Britain and Ireland?

Notes

CHAPTER 1

1. R. Muir-Wood (1989). *An Interpretation and a Compilation of the Tertiary Tectonics of Britain*, Vol. H. of UK CEGB Seismic Hazard Working Party; R. Muir-Wood, W. Aspinall, G. Woo, D. Mallard, and B.O. Skipp (1993). *The Seismotectonics of Britain*, Vol. M. of UK CEGB Seismic Hazard Working Party; R. Muir Wood (1989). 'Fifty million years of "passive margin" deformation in North West Europe.' In: S. Gregersen and P.W. Basham (eds), *Earthquakes at North Atlantic Passive Margins: Neotectonics and Postglacial Rebound*, Dordrecht: Kluwer, pp. 393–411.
2. W.A. Berggren (1998). 'The Cenozoic Era: Lyellian (chrono)stratigraphy and nomenclatural reform at the millennium.' In: D.J. Blundell and A.C. Scott (eds), *Lyell: The Past is the Key to the Present*, London: The Geological Society, Special Publications no. 143, pp. 111–132, http://sp.lyellcollection.org/content/143/1/111.full.pdf
3. F.G.H. Blyth (1962). 'The structure of the north-eastern tract of the Dartmoor Granite.' *Quarterly Journal of the Geological Society*, 118 (1–4), 435–451, https://doi.org/10.1144/gsjgs.118.1.0435
4. Wilfrid Hodges (2011). 'Heron's Brook, Sticklepath', http://wilfridhodges.co.uk/personal04.html
5. Richard Blagden, Martin Broadbent, Annie Kitto, Gina Little, and Trevor Ryder (2015). 'Geology of Chapel Hill, Torquay', http://ougs.org/files/swe/reports/Chapel_Hill_Field_Guide.pdf
6. M. Sugan, J.E.L. Wu, and K. McClay (2014). '3D analogue modelling of transtensional pull-apart basins: comparison with the Cinarcik basin, Sea of Marmara, Turkey.' *Bollettino di Geofisica Teorica ed Applicata*, 55 (4), 699–716.
7. E. Altunel, M. Meghraoui, H. Serdar Akyuz, and A. Dikbaş (2004). 'Characteristics of the 1912 co-seismic rupture along the North Anatolian Fault Zone (Turkey): implications for the expected Marmara earthquake.' *Terra Nova*, 16 (4), 198–204.

CHAPTER 2

1. John Betjeman (2008). *First and Last Loves*, London: Faber & Faber (originally published 1952).
2. Ian West, 'Alum Bay and the Needles', *Geology of the Isle of Wight*, http://www.southampton.ac.uk/~imw/Alum-Bay.htm
3. D.J. Evans, G.A. Kirby, and A.G. Hulbert (2011). 'New insights into the structure and evolution of the Isle of Wight Monocline.' *Proceedings of the Geologists' Association*, 122 (5), 764–780, http://nora.nerc.ac.uk/16024/1/Evans_et_al_PGEOLA-D-11-00019R1.pdf
4. 'John Hooke's house and its environs', *Isle of Wight History Centre*, http://www.iwhistory.org.uk/RM/hooke/house.htm
5. Albert V. Carozzi (1970). 'Robert Hooke, Rudolf Erich Raspe, and the concept of "earthquakes".' *Isis*, 61 (1), 85–91, http://www.journals.uchicago.edu/doi/abs/10.1086/350580?journalCode=isis
6. Robert Hooke (1705). Posthumous publication, *Lectures and Discourses on Earthquakes and Subterraneous Eruptions*.
7. John Phillips (1844). *Memoirs of William Smith LLD*, London: John Murray, p. 8.
8. Carlo Doglioni, Eugenio Carminati, Mattia Crespi, Marco Cuffaro, Mattia Penati, and Federica Riguzzi (2015). 'Tectonically asymmetric Earth: from net rotation to polarized westward drift of the lithosphere.' *Geoscience Frontiers*, 6 (3), 401–418, https://www.sciencedirect.com/science/article/pii/S1674987114000231
9. James D. Hays and Walter C. Pitman III (1973). 'Lithospheric plate motion, sea level changes and climatic and ecological consequences.' *Nature*, 246 (5427), 18–22, https://doi.org/10.1038/246018a0
10. P.M. Hopson (2005). 'A stratigraphical framework for the Upper Cretaceous Chalk of England and Scotland, with statements on the Chalk of Northern Ireland and the UK Offshore Sector', British Geological Survey Research Report RR/05/01, http://nora.nerc.ac.uk/3230/1/RR05001.pdf
11. P.T. Walsh (1966). 'Cretaceous outliers in south-west Ireland and their implications for Cretaceous palaeogeography.' *Quarterly Journal of the Geological Society of London*, 122 (1–4), 63–84, https://jgs.lyellcollection.org/content/122/1-4/63
12. Basil Charles King (1954). 'The Ard Bheinn area of the Central Igneous Complex of Arran.' *Quarterly Journal of the Geological Society of London*, 110 (1–4), 323–355, https://doi.org/10.1144/GSL.JGS.1954.110.01-04.15; see also A. McKirdy, J. Gordon, and R. Crofts (2007). *Land of Mountain and*

Flood: The Geology and Landforms of Scotland. Birlinn Ltd. ISBN 978-1841583570.

13. P.I. Premović, B.S. Ilić, and M.G. Đorđević (2012). 'Iridium anomaly in the cretaceous-paleogene boundary at Højerup (Stevns Klint, Denmark) and Woodside Creek (New Zealand): the question of an enormous proportion of extraterrestrial component.' *Journal of the Serbian Chemical Society*, 77 (2), 247–255, https://doi.org/10.2298/JSC110404178P
14. F. Surlyk, T. Damholt, and M. Bjerager (2006). 'Stevn's Klint, uppermost Maastrichtian chalk, Cretaceous–Tertiary boundary and lower Danian bryozoan mound complex.' *Bulletin of the Geological Society of Denmark*, 54, 1–48, http://2dgf.dk/xpdf/bull54.pdf
15. L.W. Alvarez, W. Alvarez, F. Asaro, and H.V. Michel (1980). 'Extraterrestrial cause for the Cretaceous-Tertiary extinction.' *Science*, 208 (4448), 1095–1108, https://doi.org/10.1126/science.208.4448.1095
16. R.N. Thompson and S.A. Gibson (2000). 'Transient high temperatures in mantle plume heads inferred from magnesian olivines in Phanerozoic picrites.' *Nature*, 407 (6803), 502–506, https://www.nature.com/articles/35035058
17. S.P. Holford, J.P. Turner, and P.F. Green (2005). 'Reconstructing the Mesozoic–Cenozoic exhumation history of the Irish Sea basin system using apatite fission track analysis and vitrinite reflectance data.' In: A.G. Doré and B.A. Vining (eds), *Petroleum Geology: North-West Europe and Global Perspectives—Proceedings of the 6th Petroleum Geology Conference*, London: Geological Society, pp. 1095–1107, http://www.geotrack.com.au/papers/holford_et_al_2005a.pdf
18. M. Rider and D. Kroon (2003). 'Redeposited chalk hydrocarbon reservoirs of the North Sea caused by the Chicxulub K-T bolide impact.' *Netherlands Journal of Geosciences*, 82 (4), 333–337, https://doi.org/10.1017/S0016774600020163
19. L.M. Mackay, J. Turner, S.M. Jones, and N.J. White (2005). 'Cenozoic vertical motions in the Moray Firth Basin associated with initiation of the Iceland Plume.' *Tectonics*, 24 (5), https://doi.org/10.1029/2004TC001683
20. M.J. Hole, J.M. Millett, N.W. Rogers, and D.W. Jolley (2015). 'Rifting and mafic magmatism in the Hebridean Basins.' *Journal of the Geological Society*, 172 (2), 218–236, https://doi.org/10.1144/jgs2014-100, http://aura.abdn.ac.uk/bitstream/handle/2164/5582/6957_2_merged_1414431604.pdf?sequence=1; B. Wawerzinek, J.R.R. Ritter, M. Jordan, and M. Landes (2008). 'An upper-mantle upwelling underneath Ireland revealed from non-linear tomography.' *Geophysical Journal International*, 175 (1), 253–268, https://doi.org/10.1111/j.1365-246X.2008.03908.x

21. John F. Rudge, Max E. Shaw Champion, Nicky White, Dan McKenzie, and Bryan Lovell (2008). 'A plume model of transient diachronous uplift at the Earth's surface.' *Earth and Planetary Science Letters*, 267 (1–2), 146–160, https://doi.org/10.1016/j.epsl.2007.11.040
22. B. Lovell (2010). 'A pulse in the planet: regional control of high-frequency changes in relative sea level by mantle convection.' *Journal of the Geological Society*, 167 (4), 637–648, https://doi.org/10.1144/0016-76492009-127
23. Mackay et al., 'Cenozoic vertical motions'.
24. (To be found in the Murchison papers held by the Geological Society of London.)
25. C. Lewis (2011). 'Holmes's First Date', *Geoscientist Online*, https://www.geolsoc.org.uk/Geoscientist/Archive/June-2011/Holmess-first-date

CHAPTER 3

1. James Boswell (1785), *The Journal of a Tour to the Hebrides with Samuel Johnson, LL.D.*
2. J.A. MacCulloch (1905), *The Misty Isle of Skye: Its Scenery, Its People, Its Story*, O Anderson & Ferrier, p. 320.
3. The Autobiography of Charles Darwin published in 1887 by John Murray as part of *The Life and Letters of Charles Darwin*, including an autobiographical chapter.
4. In a letter to William Whewell in March 1837.
5. A.E. Mussett, P. Dagley, and R.R. Skelhorm (1988). 'Time and duration of igneous activity in the British Tertiary Igneous Province.' In: A.C. Morton and L.M. Parsons (eds), *Early Tertiary Volcanism and the Opening of the North Atlantic*, London: The Geological Society, Special Publications no. 39, pp. 271–281.
6. M.J. Hole, J.M. Millett, N.W. Rogers, and D.W. Jolley (2015). 'Rifting and mafic magmatism in the Hebridean Basins.' *Journal of the Geological Society*, 172 (2), 218–236, https://doi.org/10.1144/jgs2014-100, http://aura.abdn.ac.uk/bitstream/handle/2164/5582/6957_2_merged_1414431604.pdf; Ray W. Kent and J. Godfrey Fitton (2000). 'Mantle sources and melting dynamics in the British Palaeogene Igneous Province.' *Journal of Petrology*, 41 (7), 1023–1040, https://doi.org/10.1093/petrology/41.7.1023
7. C.H. Emeleus and V.R. Troll (2014). 'The Rum Igneous Centre, Scotland.' *Mineralogical Magazine*, 78 (4), 805–839, http://minmag.geoscienceworld.org/content/78/4/805
8. Tobias Mattsson (2014). 'The Roots of a Magma Chamber, the Central Intrusion, Rum, NW-Scotland', Examensarbete vid Institutionen för

geovetenskaper Nr 304, Department of Earth Sciences, Uppsala University, http://www.diva-portal.org/smash/get/diva2:770853/FULLTEXT01.pdf

9. Valentin R. Troll, C. Henry Emeleus, and Colin H. Donaldson (2000). 'Caldera formation in the Rum Central Igneous Complex, Scotland.' *Bulletin of Volcanology*, 62 (4–5), 301–317, https://link.springer.com/article/10.1007/s004450000099
10. Valentin R. Troll, Graeme R. Nicoll, Colin H. Donaldson, and Henry C. Emeleus (2008). 'Dating the onset of volcanism at the Rum Igneous Centre, NW Scotland.' *Journal of the Geological Society*, 165 (3), 651–659, https://doi.org/10.1144/0016-76492006-190; E.P. Holohan, V.R. Troll, M. Errington, C.H. Donaldson, G.R. Nicoll, and C.H. Emeleus (2009). 'The Southern Mountains Zone, Isle of Rum, Scotland: volcanic and sedimentary processes upon an uplifted and subsided magma chamber roof.' *Geological Magazine*, 146 (3), 400–418, https://doi.org/10.1017/S0016756808005876
11. C.O. Ofoegbu and M.H.P. Bott (1985). 'Interpretation of the Minch linear magnetic anomaly and of a similar feature on the shelf north of Lewis by non-linear optimization.' *Journal of the Geological Society*, 142 (6), 1077–1087, https://jgs.lyellcollection.org/content/142/6/1077
12. Kent and Fitton, 'Mantle sources and melting dynamics'.
13. E.B. Bailey, C.T. Clough, W.B. Wright, J.E. Richey, and G.V. Wilson (1924). *Tertiary and Post-Tertiary Geology of Mull, Loch Aline and Oban* (Memoir of the Geological Survey of Great Britain, Scotland), Edinburgh: HMSO.
14. Colin Baxter and Jim Crumley (1998). *St Kilda: A Portrait of Britain's Remotest Island Landscape*, Biggar: Colin Baxter Photography.
15. R.W.H. Butler and D.H.W. Hutton (1994). 'Basin structure and Tertiary magmatism on Skye, NW Scotland.' *Journal of the Geological Society*, 151 (6), 931–944, https://doi.org/10.1144/gsjgs.151.6.0931
16. Michael J. Simms (2000). 'The sub-basaltic surface in northeast Ireland and its significance for interpreting the Tertiary history of the region.' *Proceedings of the Geologists' Association*, 111 (4), 321–336, https://doi.org/10.1016/S0016-7878(00)80088-7
17. David W. Jolley, Brian R. Bell, Ian T. Williamson, and Iain Prince (2009). 'Syn-eruption vegetation dynamics, paleosurfaces and structural controls on lava field vegetation: an example from the Palaeogene Staffa, Mull Lava Field, Scotland.' *Review of Palaeobotany and Palynology*, 153 (1–2), 19–33, http://www.sciencedirect.com/science/article/pii/S0034666708000924
18. Ray W. Kent, Bonita A. Thomson, Raymond R. Skelhorn, Andrew C. Kerr, Mike J. Norry, and J. Nick Walsh (1998). 'Emplacement of

Hebridean Tertiary flood basalts: evidence from an inflated pahoehoe lava flow on Mull, Scotland.' *Journal of the Geological Society*, 155 (4), 599–607, https://doi.org/10.1144/gsjgs.155.4.0599

19. Jón Steingrímsson and Keneva Kunz (1998). *Fires of the Earth: The Laki Eruption, 1783–4*, Reykjavik: University of Iceland Press.
20. James Delbourgo (2017). *Collecting the World: Hans Sloane and the Origins of the British Museum*, Cambridge, MA: Belknap Press/Harvard University Press. Reviewed by Jenny Uglow (2017). 'Collecting for the glory of God.' *The New York Review*, http://www.nybooks.com/articles/2017/10/12/hans-sloane-collecting-glory-god/
21. Thomas Pennant's *A Tour in Scotland, and Voyage to the Hebrides 1772* (published 1774–1776); Ralph Crane and Lisa Fletcher (2015). 'Inspiration and spectacle: the case of Fingal's Cave in nineteenth-century art and literature.' *ISLE: Interdisciplinary Studies in Literature and Environment*, 22 (4), 778–800, https://doi.org/10.1093/isle/isv052
22. Philip Ball (2012). 'Pattern formation in nature: physical constraints and self-organising characteristics.' *Architectural Design*, 82 (2), 22–27.
23. Ian T. Williamson and Brian R. Bell (2012). 'The Staffa Lava Formation: graben-related volcanism, associated sedimentation and landscape character during the early development of the Palaeogene Mull Lava Field, NW Scotland.' *Scottish Journal of Geology*, 48 (1), 1–46, https://doi.org/10.1144/0036-9276/01-439; J.C. Phillips, M.C.S. Humphreys, K.A. Daniels, R.J. Brown, and F. Witham (2013). 'The formation of columnar joints produced by cooling in basalt at Staffa, Scotland.' *Bulletin of Volcanology*, 75 (6): art. no. 715, http://dro.dur.ac.uk/14219/1/14219.pdf
24. Paul Lyle (2000). 'The eruption environment of multi-tiered columnar basalt lava flows.' *Journal of the Geological Society*, 157 (4), 715–722, https://doi.org/10.1144/jgs.157.4.715
25. M.A. Hamilton, D.G. Pearson, R.N. Thompson, S.P. Kelley, and C.H. Emeleus (1998). 'Rapid eruption of Skye lavas inferred from precise U–Pb and Ar–Ar dating of the Rum and Cuillin plutonic complexes.' *Nature*, 394 (6690), 260–263, http://www.nature.com/nature/journal/v394/n6690/full/394260a0.html
26. Bailey et al., *Tertiary and Post-Tertiary Geology*; Andrew C. Kerr, Ray W. Kent, Bonita A. Thomson, Jon K. Seedhouse, and Colin H. Donaldson (1999). 'Geochemical evolution of the Tertiary Mull Volcano, western Scotland.' *Journal of Petrology*, 40 (6), 873–908, https://doi.org/10.1093/petroj/40.6.873
27. Simon P. Holford, Paul F. Green, Richard R. Hillis, Jonathan P. Turner, and Carl T.E. Stevenson (2009). 'Mesozoic–Cenozoic exhumation and

volcanism in Northern Ireland constrained by AFTA and compaction data from the Larne No. 2 borehole.' *Petroleum Geoscience*, 15 (3), 239–257, https://doi.org/10.1144/1354-079309-840

28. D. Wilcox, 'Rendezvous with a swarm of oxymoronic priapic dykes on the SW coast of Jura', https://seakayakphoto.blogspot.co.uk/2016/05/rendezvous-with-swarm-of-oxymoronic.html
29. C.H. Emeleus and B.R. Bell (2005). 'Dykes, dyke-swarms and volcanic plugs, Palaeogene volcanic districts of Scotland.' In: *British Regional Geology: The Palaeogene Volcanic Districts of Scotland*, fourth edition, Keyworth, Nottingham: British Geological Survey, http://earthwise.bgs.ac.uk/index.php/Dykes,_dyke_swarms_and_volcanic_plugs,_Palaeogene_volcanic_districts_of_Scotland
30. K. Saemundsson (1979). 'Outline of the geology of Iceland.' *Jokull*, 29, 7–27.
31. H. Sigurdsson and S.R.J. Sparks (1978). 'Lateral magma flow within rifted Icelandic crust.' *Nature*, 274 (5667), 126–130, https://doi.org/10.1038/274126a0
32. O. Sigurdsson (1980). 'Surface deformation of the Krafla Fissure Swarm in two rifting events.' *Journal of Geophysics*, 47, 154–159.
33. W. Roger Buck, Páll Einarsson, and Bryndís Brandsdóttir (2006). 'Tectonic stress and magma chamber size as controls on dike propagation: constraints from the 1975–1984 Krafla rifting episode.' *Journal of Geophysical Research*, 111 (B12), https://doi.org/10.1029/2005JB003879
34. P. Einarsson and B. Brandsdottir (1980). 'Seismological evidence for lateral magma intrusion during the July 1978 deflation of the Krafla volcano in NE-Iceland.' *Journal of Geophysics*, 47, 160–165.
35. G.P.L. Walker (1959). 'Some observations on the Antrim basalts and associated dolerite intrusions.' *Proceedings of the Geologists' Association*, 70 (2), 179–205, https://www.sciencedirect.com/science/article/abs/pii/S0016787859800031
36. Morgane Ledevin, Nicholas Arndt, Mark R. Cooper, Garth Earls, Paul Lyle, Charles Aubourg, and Eric Lewin (2012). 'Intrusion history of the Portrush Sill, County Antrim, Northern Ireland: evidence for rapid emplacement and high-temperature contact metamorphism.' *Geological Magazine*, 149 (1), 67–79, https://doi.org/10.1017/S0016756811000537
37. Fergus G.F. Gibb and Sally Gibson (1989). 'The Little Minch Sill Complex.' *Scottish Journal of Geology*, 25 (3), 367–370, https://doi.org/10.1144/sjg25030367; C.H. Emeleus and B.R. Bell (2005). 'Sills and sill-complexes, Palaeogene volcanic districts of Scotland.' In: *British Regional Geology: The Palaeogene Volcanic Districts of Scotland*, fourth edition,

Keyworth, Nottingham: British Geological Survey, http://earthwise.bgs.ac.uk/index.php/Sills_and_sill-complexes,_Palaeogene_volcanic_districts_of_Scotland

38. Brian R. Bell and Ian T. Williamson (2013). 'Palaeocene intracanyon-style lava emplaced during the early shield-building stage of the Cuillin Volcano, Isle of Skye, NW Scotland.' *Earth and Environmental Science Transactions of The Royal Society of Edinburgh*, 104 (2), 205–230, https://doi.org/10.1017/S1755691013000509
39. Christian Huber, Meredith Townsend, Wim Degruyter, and Olivier Bachmann (2019). 'Optimal depth of subvolcanic magma chamber growth controlled by volatiles and crust rheology.' *Nature Geoscience*, 12 (9), 762–768, https://doi.org/10.1038/s41561-019-0415-6; Brown University (2019). 'Why there's a "sweet spot" depth for underground magma chambers', *ScienceDaily*, 19 August, https://www.sciencedaily.com/releases/2019/08/190819132125.htm
40. Walker, 'Some observations on the Antrim basalts'.
41. C.T.E. Stevenson and N. Bennett (2011). 'The emplacement of the Palaeogene Mourne Granite Centres, Northern Ireland: new results from the Western Mourne Centre.' *Journal of the Geological Society*, 168 (4), 831–836, https://doi.org/10.1144/0016-76492010-123
42. P.J. Gibson and P. Lyle (1993). 'Evidence for a major Tertiary dyke swarm in County Fermanagh, Northern Ireland, on digitally processed aeromagnetic imagery.' *Journal of the Geological Society*, 150 (1), 37–38, https://doi.org/10.1144/gsjgs.150.1.0037
43. B.C. Chacksfield (2010). 'A preliminary interpretation of Tellus airborne magnetic and electromagnetic data for Northern Ireland', British Geological Survey Internal Report, IR/07/041, https://core.ac.uk/download/pdf/57534.pdf

CHAPTER 4

1. Tyler K. Ambrose and Michael P. Searle (2019). '3-D structure of the Northern Oman-UAE ophiolite: widespread, short-lived, suprasubduction zone magmatism.' *Tectonics*, 38 (1), 233–252, https://doi.org/10.1029/2018TC005038
2. C.H. Emeleus and B.R. Bell (2005). 'Dykes, dyke-swarms and volcanic plugs, Palaeogene volcanic districts of Scotland.' In: *British Regional Geology: The Palaeogene Volcanic Districts of Scotland*, fourth edition, Keyworth, Nottingham: British Geological Survey, http://earthwise.bgs.ac.uk/index.php/Dykes,_dyke_swarms_and_volcanic_plugs,_

Palaeogene_volcanic_districts_of_Scotland; John Michael Speight (1972). 'The form and structure of the Tertiary dyke-swarms of Skye and Ardnamurchan', Doctor of Philosophy thesis, University of London, https://kclpure.kcl.ac.uk/portal/files/2929693/318998.pdf; R.J.H. Jolly and D.J. Sanderson (1995). 'Variation in the form and distribution of dykes in the Mull swarm.' *Scotland Journal of Structural Geology*, 17 (11), 1543–1557, https://www.sciencedirect.com/science/article/pii/019181419500046G

3. Sarah J. Fowler, Wendy A. Bohrson, and Frank J. Spera (2004). 'Magmatic evolution of the Skye Igneous Centre, Western Scotland: modelling of assimilation, recharge and fractional crystallization.' *Journal of Petrology*, 45 (12), 2481–2505, https://academic.oup.com/petrology/article/45/12/2481/1545372
4. G.W. Tyrrell (1928). *The Geology of Arran*. Memoirs of the Geological Survey, Scotland. Edinburgh: His Majesty's Stationery Office.
5. G.P.L. Walker (1959). 'Some observations on the Antrim basalts and associated dolerite intrusions.' *Proceedings of the Geologists' Association*, 70 (2), 179–205, https://www.sciencedirect.com/science/article/abs/pii/S0016787859800031
6. A.C. Morton (1982). 'Lower Tertiary sand development in the Viking Graben, North Sea.' *Bulletin of the American Association of Petroleum Geologists*, 66 (10), 1542–1559.
7. P. Einarsson (1991). 'Earthquakes and present-day tectonism in Iceland.' *Tectonophysics*, 189 (1–4), 261–279, https://doi.org/10.1016/0040-1951(91)90501-I
8. Emeleus and Bell, 'Dykes, dyke-swarms and volcanic plugs'.
9. N. Holgate (1969). 'Palaeozoic and Tertiary transcurrent fault movements on the Great Glen Fault.' *Scottish Journal of Geology*, 5 (2), 97–139, https://doi.org/10.1144/sjg05020097; M. Bacon and J. Chesher (1974). 'Evidence against post-Hercynian transcurrent movement on the Great Glen Fault.' *Scottish Journal of Geology*, 11, 79–82, https://doi.org/10.1144/sjg11010079
10. R.J. Walker, R.E. Holdsworth, J. Imber, and D. Ellis (2011). 'Onshore evidence for progressive changes in rifting directions during continental break-up in the NE Atlantic.' *Journal of the Geological Society*, 168 (1), 27–48, https://doi.org/10.1144/0016-76492010-021; David Ellis and Martyn S. Stoker (2014). 'The Faroe–Shetland Basin: a regional perspective from the Paleocene to the present day and its relationship to the opening of the North Atlantic Ocean.' In: S.J.C. Cannon and D. Ellis (eds), *Hydrocarbon Exploration to Exploitation West of Shetlands*, London: The Geological Society, Special Publications no. 397, pp. 11–31,

http://nora.nerc.ac.uk/id/eprint/503195/1/Ellis%20and%20Stoker_SP397_2014.pdf

11. Søren B. Nielsen, Randell Stephenson, and Erik Thomsen (2008). 'Dynamics of Mid-Palaeocene North Atlantic rifting linked with European intra-plate deformations.' *Nature*, 450 (7172), 1071–1074, https://doi.org/10.1038/nature06379
12. M.R. Cooper, H. Anderson, J.J. Walsh, C.L. van Dam, M.E. Young, G. Earls, and A. Walker (2012). 'Palaeogene Alpine tectonics and Icelandic plume-related magmatism and deformation in Northern Ireland.' *Journal of the Geological Society*, 169 (1), 29–36, http://nora.nerc.ac.uk/id/eprint/16421/1/Cooperetal_Tellusdykes_JGS_accepted_version_text_only.pdf
13. B.C. Chacksfield (2010). 'A preliminary interpretation of Tellus airborne magnetic and electromagnetic data for Northern Ireland', British Geological Survey Internal Report, IR/07/041, https://core.ac.uk/download/pdf/57534.pdf
14. M.P. Tate and M.R. Dobson (1989). 'Late Permian to early Mesozoic rifting and sedimentation offshore NW Ireland.' *Marine and Petroleum Geology*, 6 (1), 49–59.
15. D. Naylor and P.M. Shannon (2005). 'The structural framework of the Irish Atlantic Margin.' *Geological Society, London, Petroleum Geology Conference series*, 6, 1009–1021, https://doi.org/10.1144/0061009
16. Nicky White, Michael Tate, and John-Joe Conroy (1992). 'Lithospheric stretching in the Porcupine Basin, west of Ireland.' In: J. Parnell (ed.), *Basins on the Atlantic Seaboard: Petroleum Geology, Sedimentology and Basin Evolution*, London: The Geological Society, Special Publications no. 62, pp. 327–349.
17. Morgan Ganerød, Fiona Meade, Valentin Troll, Henry Emeleus, and David Chew (2013). 'Major felsic volcanism during the interbasaltic "quiet period" of the Antrim plateau basalts', id. EGU2013-7119, EGU General Assembly 2013, 7–12 April, Vienna, Austria, http://adsabs.harvard.edu/abs/2013EGUGA..15.7119G
18. Michael Krumbholz, Christoph F. Hieronymus, Steffi Burchardt, Valentin R. Troll, David C. Tanner, and Nadine Friese (2014). 'Weibull-distributed dyke thickness reflects probabilistic character of host-rock strength.' *Nature Communications*, 5: art. no. 3272, https://www.nature.com/articles/ncomms4272
19. 'Dykes and sills', *Geology North*, http://www.geologynorth.uk/?page_id=15468

20. I.B. Cameron and D. Stephenson (1985). 'Tertiary igneous intrusions, Midland Valley of Scotland.' In: *British Regional Geology: The Midland Valley of Scotland*, third edition, Keyworth, Nottingham: British Geological Survey, http://earthwise.bgs.ac.uk/index.php/Tertiary_igneous_intrusions,_Midland_Valley_of_Scotland
21. 'Tynemouth Dyke', *Geology North*, http://www.geologynorth.uk/?page_id=18066
22. 'Acklington Dyke', *Geology North*, http://www.geologynorth.uk/?page_id=18832
23. R. Macdonald, L. Wilson, R.S. Thorpe, and A. Martin (1988). 'Emplacement of the Cleveland Dyke: evidence from geochemistry, mineralogy, and physical modelling.' *Journal of Petrology*, 29 (3), 559–583, https://doi.org/10.1093/petrology/29.3.559
24. Whinstone Mining (extracts from chapters of *Great Ayton: A History of the Village*, by Dan O'Sullivan), https://greatayton.wdfiles.com/local--files/whinstone/Whinstone-Mining.pdf
25. Mostyn Wall, Joe Cartwright, Richard Davies, and Andrew McGrandle (2010). '3D seismic imaging of a Tertiary Dyke Swarm in the Southern North Sea, UK.' *Basin Research*, 22 (2), 181–194, https://doi.org/10.1111/j.1365-2117.2009.00416.x; Marta Swierczek, John R. Underhill, Helen Lever, and David Millward (2010). 'Role of the Base Permian Unconformity in controlling Carboniferous reservoir prospectivity, UK southern North Sea', Poster Three, Search and Discovery Article #10275, 22 November, adapted from poster presentation at AAPG Annual Convention and Exhibition, New Orleans, LA, 11–14 April, http://www.searchanddiscovery.com/pdfz/documents/2010/10275swierczek/poster03.pdf.html
26. M. Kortekaas, U. Böker, C. Van Der Kooij, and B. Jaarsma (2018). 'Figure 9, Lower Triassic reservoir development in the northern Dutch offshore.' In: B. Kilhams, P.A. Kukla, S. Mazur, T. Mckie, H.F. Mijnlieff, and K. Van Ojik (eds), *Mesozoic Resource Potential in the Southern Permian Basin*, London: The Geological Society, Special Publications no. 469, pp. 149–168, https://doi.org/10.1144/SP469.19, https://sp.lyellcollection.org/content/specpubgsl/early/2018/04/12/SP469.19.full.pdf; D.B. Thompson and J.A. Winchester (1995). 'Field relationships, geochemistry, and tectonic context of the Tertiary dyke suites in Staffordshire and Shropshire, central England.' *Proceedings of the Yorkshire Geological Society*, 50 (3), 191–208, https://doi.org/10.1144/pygs.50.3.191

27. John F. Dewey (1982). 'Plate tectonics and the evolution of the British Isles, Thirty-fifth William Smith Lecture.' *Journal of the Geological Society*, 139 (4), 371–412, https://doi.org/10.1144/gsjgs.139.4.0371
28. S.R. Kirton and J.A. Donato (1985). 'Some buried Tertiary dykes of Britain and surrounding waters deduced by magnetic modelling and seismic reflection methods.' *Journal of the Geological Society*, 142 (6), 1047–1057, https://doi.org/10.1144/gsjgs.142.6.1047
29. A. Harker (1887–8). 'Woodwardian Museum Notes: on some Anglesey Dykes.' *Geological Magazine*, parts I, II, 1887, 4 (9), 409–416 and 4 (12), 546–552; part III, 1888, 5 (6), 267–272.
30. Richard E. Bevins, Jana M. Horák, A.D. Evans, and R. Morgan (1996). 'Palaeogene dyke swarm, NW Wales: evidence for Cenozoic sinistral fault movement.' *Journal of the Geological Society*, 153 (2), 177–180, https://doi.org/10.1144/gsjgs.153.2.0177
31. D. Williams (1930). 'The geology of the country between Nant Peris and Nant Ffrancon, Snowdonia.' *Quarterly Journal of the Geological Society*, 86 (1–4), 191–232; D. Williams (1924). 'On two olivine-dolerite dykes in Snowdonia.' *Proceedings of the Liverpool Geological Society*, 14, 38–47; Edward Greenly (1937). 'The age of the mountains of Snowdonia.' In: *The Geology of Snowdonia: A Collection of Historical Articles on the Physical Features of the Peaks of Snowdonia*, Read Books Ltd, 2011.
32. A.A. Archer and R.W. Elliot (1965). 'The occurrence of olivine-dolerite dykes near Llanwrst, north Wales.' *Bulletin of the Geological Survey of Great Britain*, 23, 145–152.
33. K. Hotchkiss, 'Grinshill: Mike Rosenbaum: 17th July 2013', http://www.shropshiregeology.org.uk/sgspublications/Newsletters/2013-09%20SGS%20Newsletter%2006.pdf; R.W. Pocock and D.A. Wray (1925). *The Geology of the Country around Wem*. Memoirs of the Geological Survey of England and Wales. Explanation of Sheet 138.
34. D.G. Quirk and G.S. Kimbell (1997). 'Structural evolution of the Isle of Man and central part of the Irish Sea.' In: N.S. Meadows, S.P. Trueblood, M. Hardman, and G. Cowan (eds), *Petroleum Geology of the Irish Sea and Adjacent Areas*, London: The Geological Society, Special Publications no. 124, pp. 135–159; P. Stone, D. Millward, B. Young, J.W. Merritt, S.M. Clarke, M. McCormac, and D.J.D. Lawrence (2010). 'Late Mesozoic and Cenozoic tectonics and magmatism, northern England.' In: *British Regional Geology: Northern England*, fifth edition, Keyworth, Nottingham: British Geological Survey, http://earthwise.bgs.ac.uk/index.php/Late_Mesozoic_and_Cenozoic_tectonics_and_magmatism,_Northern_England
35. Stone et al., 'Late Mesozoic and Cenozoic tectonics'.

36. G. Arter and S.W. Fagin (1993). 'The Fleetwood Dyke and the Tynwald fault zone, Block 113/27, East Irish Sea Basin.' *Geological Society, London, Petroleum Geology Conference series*, 4, 835–843, https://doi.org/10.1144/0040835
37. David B. Thompson (1993). 'Murchison's and Darwin's dykes in North Staffordshire', British Association for the Advancement of Science, Keele. Section C: Geology Excursion notes, pp. 1–8, https://nsgga.files.wordpress.com/2019/08/1993murchisondarwindykes.pdf
38. David B. Thompson (2000). 'Charles Darwin in North Staffordshire', British Association for the Advancement of Science, Field Excursion, 26 March, https://nsgga.files.wordpress.com/2019/08/2000darwinnorthstaffs.pdf
39. W.T.C. Sowerbutts (1987). 'Magnetic mapping of the Butterton Dyke: an example of detailed geophysical surveying.' *Journal of the Geological Society*, 144 (1), 29–33, https://doi.org/10.1144/gsjgs.144.1.0029; Thompson and Winchester, 'Field relationships, geochemistry, and tectonic context'.
40. F.W. Cope and C.S. Exley (1988). 'Discussion on the magnetic mapping of the Butterton Dyke: an example of detailed geophysical surveying.' *Journal of the Geological Society*, 145 (1), 181–184, https://doi.org/10.1144/gsjgs.145.1.0181; *Memoirs of the North Staffordshire Coalfield* (1905). London: HMSO.
41. Thompson and Winchester, 'Field relationships, geochemistry, and tectonic context'.
42. N.J. Fortey (1991). 'Petrology of the Caton Dyke', British Geological Survey Technical Report WG/91/11.
43. A. Brandon (1992). 'Geology of the Littledale area, 1:10 000 sheet SD56SE, part of 1:50 000 sheet 59 (Lancaster)', Natural Environment Research Council, British Geological Survey Onshore Geology Series, Technical Report WA/92/16, http://nora.nerc.ac.uk/id/eprint/10305/1/WA92016.pdf
44. Cameron and Stephenson, 'Tertiary igneous intrusions'.
45. O. Ishizuka, R.N. Taylor, N. Geshi, and N. Mochizuki (2017). 'Large-volume lateral magma transport from the Mull Volcano: an insight to magma chamber processes.' *Geochemistry, Geophysics, Geosystems*, 18 (4), 1618–1640, https://doi.org/10.1002/2016GC006712
46. R.J.H. Jolly and D.J. Sanderson (1995). 'Variation in the form and distribution of dykes in the Mull swarm, Scotland.' *Journal of Structural Geology*, 17 (11), 1543–1557, https://www.sciencedirect.com/science/article/pii/019181419500046G; Larry G. Mastin and David D. Pollard (1988). 'Surface deformation and shallow dike intrusion processes at Inyo

Craters, Long Valley, California.' *Journal of Geophysical Research: Solid Earth*, 93 (B11), 13221–13235, https://doi.org/10.1029/JB093iB11p13221; M. Casey, C. Ebinger, D. Keir, R. Gloaguen, and F. Mohamed (2006). 'Strain accommodation in transitional rifts: extension by magma intrusion and faulting in Ethiopian rift magmatic segments.' In: G. Yirgu, C.J. Ebinger, and P.K.H. Maguire (eds), *The Afar Volcanic Province within the East African Rift System*, London: The Geological Society, Special Publications no. 259, pp. 143–163, https://doi.org/10.1144/GSL.SP.2006.259.01.13; K.A. Daniels, I.D. Bastow, D. Keir, R.S.J. Sparks, and T. Menand (2014). 'Thermal models of dyke intrusion during development of continent–ocean transition.' *Earth and Planetary Science Letters*, 385, 145–153, http://www.sciencedirect.com/science/article/pii/S0012821X13005293

47. Ganerød et al., 'Major felsic volcanism'.
48. Cooper et al., 'Palaeogene Alpine tectonics'.
49. P.J. Gibson and P. Lyle (1993). 'Evidence for a major Tertiary dyke swarm in County Fermanagh, Northern Ireland, on digitally processed aeromagnetic imagery.' *Journal of the Geological Society*, 150 (1), 37–38, https://doi.org/10.1144/gsjgs.150.1.0037
50. H. Anderson, J.J. Walsh, and M.R. Cooper (2016). 'Faults, intrusions and flood basalts: the Cenozoic structure of the north of Ireland.' In: M.E. Young (ed.), *Unearthed: Impacts of the Tellus Surveys of the North of Ireland*, Dublin: Royal Irish Academy, 179–189, https://www.ria.ie/sites/default/files/chapter_14.pdf
51. P. Mohr (1982). 'Tertiary Dolerite intrusions of west-Central Ireland.' *Proceedings of the Royal Irish Academy*, 82B, 53–82.
52. Granyia (2016). 'Volcanic Northern Ireland—the Antrim Plateau', *Volcano Hot Spot*, https://volcanohotspot.wordpress.com/2016/10/09/volcanic-northern-ireland-the-antrim-plateau/
53. R.K. Harrison, P. Stone, I.B. Cameron, R.W. Elliot, and R.R. Harding (1987). 'Geology, petrology and geochemistry of Ailsa Craig.' *Ayrshire BGS Report*, 16 (9), http://nora.nerc.ac.uk/505002/1/Ailsa%20Craig.pdf
54. Andrew Kay and Co, Ltd. (2017). 'Ailsa Craig: The World's Best Granite', http://www.kayscurling.com/ailsa-craig-granite.html
55. A.T.J. Dollar (1968). 'Tertiary dyke swarm of Lundy, Volcanic studies group meeting, 24 January.' *Proceedings of the Geological Society London*, no. 1649, pp. 119–120.
56. Tim Holland and Roger Powell (2001). 'Calculation of phase relations involving haplogranitic melts using an internally consistent thermodynamic

dataset.' *Journal of Petrology*, 42 (4), 673–683, https://doi.org/10.1093/petrology/42.4.673

57. J.-H. Charles, M.G. Whitehouse, J.C.Ø. Andersen, R.K. Shail, and M.P. Searle (2018). 'Age and petrogenesis of the Lundy Granite: Paleocene intraplate peraluminous magmatism in the Bristol Channel, UK.' *Journal of the Geological Society*, 175 (1), 44–59, https://doi.org/10.1144/jgs2017-023; C.L. Roberts and S.G. Smith (1994). 'A new magnetic survey of Lundy Island, Bristol Channel.' *Proceedings of the Ussher Society*, 8, 293–297.

CHAPTER 5

1. C.P. Green (1997). 'The provenance of the rocks used in the construction of Stonehenge.' *Proceedings of the British Academy*, 92, 257–270, https://www.thebritishacademy.ac.uk/documents/3922/92p257.pdf M.A. Summerfield and A.S. Goudie (1980). 'The sarsens of southern England: their palaeoenvironmental interpretation with reference to other silcretes.' In: D.K.C. Jones (ed.), *The Shaping of Southern England*, London: Academic Press, pp. 71–100.
2. David J. Nash, T. Jake R. Ciborowski, J. Stewart Ullyott, Mike Parker Pearson, Timothy Darvill, Susan Greaney, Georgios Maniatis, and Katy A. Whitaker (2020). 'Origins of the sarsen megaliths at Stonehenge.' *Science Advances*, 6 (31), eabc0133, https://advances.sciencemag.org/content/6/31/eabc0133
3. C.E. Burt, J.A. Aspden, J.R. Davies, M. Hall, D.I. Schofield, T.H. Sheppard, R.A. Waters, P.R. Wilby, and M. Williams (2012). 'Geology of the Fishguard district—a brief explanation of the geological map.' *Sheet Explanation of the British Geological Survey*. 1:50 000 Sheet 210 Fishguard (England and Wales).
4. Katy Whitaker (2017). 'Wiltshire Sarsen Stones—a comic', *Wiltshire & Swindon History Centre*, https://wshc.org.uk/blog/item/wiltshire-sarsen-stones.html
5. Jennifer Huggett (2016). 'Puddingstones and related silcretes of the Anglo-Paris Basin—an overview.' *Proceedings of the Geologists' Association*, 127 (3), 297–300, https://doi.org/10.1016/j.pgeola.2016.06.002
6. Bryan Lovell (2015). 'Hertfordshire Puddingstone: relationship to the Paleocene-Eocene Thermal Maximum and a perspective on human-induced climate change.' *Proceedings of the Geologists' Association*, 121 (1), 8–13, https://doi.org/10.1016/j.pgeola.2014.12.004

7. Bryan Lovell and Jane Tubb (2006). 'Ancient quarrying of rare *in situ* Palaeogene Hertfordshire Puddingstone.' *Mercian Geologist*, 16 (3), 185–189, https://web.archive.org/web/20120425152549/http://www.ehgc.org.uk/Lovell_Tubb_Mercian_2006.pdf
8. Huggett, 'Puddingstones and related silcretes'.
9. S. Thorarinsson (1981). 'Greetings from Iceland: ash-falls and volcanic aerosols in Scandinavia.' *Geografiska Annaler*, 63A, 109–117, https://doi.org/10.1080/04353676.1981.11880024
10. David W. Jolley and Brian R. Bell (2002). 'The evolution of the North Atlantic Igneous Province and the opening of the NE Atlantic rift.' In: D.W. Jolley and B.R. Bell (eds), *The North Atlantic Igneous Province: Stratigraphy, Tectonic, Volcanic and Magmatic Processes*, London: The Geological Society, Special Publications no. 197, pp. 1–13, http://sp.lyellcollection.org/content/197/1/1.full.pdf; J. Hansen, D.A. Jerram, K. McCaffrey, and S.R. Passey (2009). 'The onset of the North Atlantic Igneous Province in a rifting perspective.' *Geological Magazine*, 146 (3), 309–325, https://doi.org/10.1017/S0016756809006347; David Ellis and Martyn S. Stoker (2014). 'The Faroe–Shetland Basin: a regional perspective from the Paleocene to the present day and its relationship to the opening of the North Atlantic Ocean.' In: S.J.C. Cannon and D. Ellis (eds), *Hydrocarbon Exploration to Exploitation West of Shetlands*, London: The Geological Society, Special Publications no. 397, pp. 11–31, http://nora.nerc.ac.uk/503195/1/Ellis%20and%20Stoker_SP397_2014.pdf
11. H. Egger and E. Brückl (2006). 'Gigantic volcanic eruptions and climatic change in the early Eocene.' *International Journal of Earth Sciences*, 95 (6), 1065–1070, https://doi.org/10.1007/s00531-006-0085-7
12. J.M. Huggett and R.W.O'B. Knox (2006). 'Clay mineralogy of the Tertiary onshore and offshore strata of the British Isles.' *Clay Minerals*, 41 (1), 5–46, https://doi.org/10.1180/0009855064110195
13. Joyce Dixon (2013). 'The secret history of the London brick', *Londonist*, https://londonist.com/2013/11/the-secret-history-of-the-london-brick

CHAPTER 6

1. P. Japsen (1997). 'Regional Neogene exhumation of Britain and the western North Sea.' *Journal of the Geological Society*, 154 (2), 239–247, https://doi.org/10.1144/gsjgs.154.2.0239
2. C. Lyell (1833). *Principles of Geology*, first edition, vol. iii, p. 293.
3. G. Yielding, J.A. Jackson, G.C.P. King, H. Sinvhal, C. Vita-Finzi, and R.M. Wood (1981). 'Relations between surface deformation, fault

geometry, seismicity, and rupture characteristics during the El Asnam (Algeria) earthquake of 10 October 1980.' *Earth and Planetary Science Letters*, 56, 287–304, https://doi.org/10.1016/0012-821X(81)90135-7

4. G.C.P. King and C. Vita-Finzi (1981). 'Active folding in the Algerian earthquake of 10 October 1980.' *Nature*, 292 (5818), 22–26, https://doi.org/10.1038/292022a0
5. Ross S. Stein and Robert S. Yeats (1989). 'Hidden earthquakes.' *Scientific American*, 260 (6), 48–59.
6. H.G. Dines, F.H. Edmunds, and C.P. Chatwin (1929). *Geology of the Country around Aldershot and Guildford*. Memoirs of the Geological Survey, DF285.
7. B.H. Mottram (1961). 'Contributions to the geology of the Mere Fault and the Vale of Wardour anticline.' *Proceedings of the Geologists' Association*, 72 (2), 187–203, https://doi.org/10.1016/S0016-7878(61)80002-3
8. R.A. Chadwick, N. Kenolty, and A. Whittaker (1983). 'Crustal structure beneath southern England from deep seismic reflection profiles.' *Journal of the Geological Society*, 140 (6), 893–911, https://doi.org/10.1144/gsjgs.140.6.0893
9. J.R. Underhill and S. Paterson (1998). 'Genesis of tectonic inversion structures: seismic evidence for the development of key structures along the Purbeck–Isle of Wight disturbance.' *Journal of the Geological Society*, 155 (6), 975–992, https://doi.org/10.1144/gsjgs.155.6.0975
10. H.J. Osborne White (1921). *Geology of the Isle of Wight*. Memoirs of the Geological Survey of England and Wales.
11. A.J. Smith and D. Curry (1975). 'The structure and geological evolution of the English Channel.' *Philosophical Transactions of the Royal Society A*, 279 (1288), 3–20, https://doi.org/10.1098/rsta.1975.0036
12. Dines et al., *Geology of the Country*.
13. A.S. Gale, P.A. Jeffery, J.M. Huggett, and P. Connolly (1999). 'Eocene inversion history of the Sandown Pericline, Isle of Wight, southern England.' *Journal of the Geological Society*, 156 (2), 327–339, https://doi.org/10.1144/gsjgs.156.2.0327
14. A.G. Plint (1982). 'Eocene tectonics and sedimentation in the Hampshire Basin.' *Journal of the Geological Society*, 139 (3), 249–254, https://doi.org/10.1144/gsjgs.139.3.0249
15. P. Ayarza, J.R. Martínez Catalán, J. Alvarez-Marrón, H. Zeyen, and C. Juhlin (2004). 'Geophysical constraints on the deep structure of a limited ocean-continent subduction zone at the North Iberian Margin.' *Tectonics*, 23 (1), https://doi.org/10.1029/2002TC001487

16. J. Alvarez-Marron, E. Rubio, and M. Torne (1997). 'Subduction related structures in the North Iberian Margin.' *Journal of Geophysical Research: Solid Earth*, 102 (B10), 22497–22511, https://doi.org/10.1029/97JB01425
17. A. Blondeau, Claude Cavelier, and Charles Pomerol (1964). 'Influence de la tectonique du Pays de Bray sur les formations paleogenes au voisinage de sa terminaison orientale.' *Bulletin de la Société Géologique de France*, S7-VI (3), 357–367, https://doi.org/10.2113/gssgfbull.S7-VI.3.357
18. Randall R. Parrish, Claire M. Parrish, and Stephanie Lasalle (2018). 'Vein calcite dating reveals Pyrenean orogen as cause of Paleogene deformation in southern England.' *Journal of the Geological Society*, 175 (3), 425–442, https://jgs.lyellcollection.org/content/175/3/425
19. D.G. Masson, J.A. Cartwright, L.M. Pinheiro, R.B. Whitmarsh, M.-O. Beslier, and H. Roeser (1994). 'Compressional deformation at the ocean–continent transition in the NE Atlantic.' *Journal of the Geological Society*, 151 (4), 607–613, https://doi.org/10.1144/gsjgs.151.4.0607
20. D.G. Masson and L.M. Parson (1984). 'Eocene deformation on the continental margin SW of the British Isles.' *Journal of the Geological Society*, 140 (6), 913–920, https://doi.org/10.1144/gsjgs.140.6.0913
21. M.J.M. Cunningham, A.L. Densmore, P.A. Allen, W.E.A. Philipps, S.D. Bennett, K. Gallagher, and A. Carter (2003). 'Evidence for post-early Eocene activity in southeastern Ireland.' *Geological Magazine*, 140 (2), 101–118, https://doi.org/10.1017/S0016756802007240
22. R.M. Tucker and G. Arter (1987). 'The tectonic evolution of the North Celtic Sea and Cardigan Bay basins with special reference to basin inversion.' *Tectonophysics*, 137 (1–4), 291–307, https://doi.org/10.1016/0040-1951(87)90324-6
23. A. Ruffell and P.F. Carey (2001). 'The northwestward continuation of the Sticklepath Fault: Bristol Channel, SW Wales, St. George's Channel and Ireland.' *Geoscience in South-West England*, 10, 134–141, https://ussher.org.uk/wp-content/uploads/journal/2001/03-Ruffell_Carey_2001.pdf
24. Pablo Rodriguez-Salgado, C. Childs, P.M. Shannon, and John J. Walsh (2017). 'Structural controls on different styles of Cenozoic inversion in the Celtic Sea Basins, offshore Ireland.' In: *Conference Proceedings, 79th EAGE Conference and Exhibition 2017*, June, pp. 1–5, https://doi.org/10.3997/2214-4609.201701288
25. S.P. Srivastava, H. Schouten, W.R. Roest, K.D. Klitgord, L.C. Kovacs, J. Verhoef, and R. Macnab (1990). 'Iberian plate kinematics: a jumping plate boundary between Eurasia and Africa.' *Nature*, 344 (6268), 756–759, https://doi.org/10.1038/344756a0

26. Gareth Williams, Jonathan P. Turner, and Simon Holford (2005). 'Inversion and exhumation of the St. George's Channel Basin, offshore Wales, UK, January 2005.' *Journal of the Geological Society*, 162 (1), 97–110, https://doi.org/10.1144/0016-764904-023, http://nora.nerc.ac.uk/id/eprint/18234/1/SGCB.pdf
27. M.F. Howells (2007). 'Palaeogene–Neogene of Wales.' In: *British Regional Geology: Wales*, Keyworth, Nottingham: British Geological Survey, http://earthwise.bgs.ac.uk/index.php/Palaeogene%E2%80%94Neogene_of_Wales
28. G.M. Dunford, P.M. Dancer, and K.D. Long (2001). 'Hydrocarbon potential of the Kish Bank Basin: integration within a regional model for the Greater Irish Sea Basin.' In: D. Corcoran, D.W. Haughton, and P.M. Shannon (eds), *The Petroleum Exploration of Ireland's Offshore Basins*, London: The Geological Society, Special Publications no. 188, pp. 135–155.
29. Hugh Anderson, John J. Walsh, and Mark R. Cooper (2018). 'The development of a regional-scale intraplate strike-slip fault system: Alpine deformation in the north of Ireland.' *Journal of Structural Geology*, 116, 47–63, https://doi.org/10.1016/j.jsg.2018.07.002
30. M.F. Quinn (2006). 'Lough Neagh: the site of a Cenozoic pull-apart basin.' *Scottish Journal of Geology*, 42 (2), 101–112, https://doi.org/10.1144/sjg42020101, http://nora.nerc.ac.uk/id/eprint/18825/1/Quinn_2006_Lough_Neagh.pdf

CHAPTER 7

1. J.F. Dewey and B.F. Windley (1988). 'Palaeocene-Oligocene tectonics of NW Europe.' In: A.C. Morton and L.M. Parsons (eds), *Early Tertiary Volcanism and the Opening of the N E Atlantic*, London: The Geological Society, Special Publications no. 39, pp. 25–31.
2. Thierry Villemin, Francis Alvarez, and Jacques Angelier (1986). 'The Rhinegraben: extension, subsidence and shoulder uplift.' *Tectonophysics*, 128 (1–2), 47–59, https://www.sciencedirect.com/science/article/pii/0040195186903070
3. B. Van Hoorn (1987). 'Structural evolution, timing and tectonic style of the Sole Pit inversion.' *Tectonophysics*, 137 (1–4), 239–254, 259–268, 270–284, https://www.sciencedirect.com/science/article/pii/0040195187903222; Martyn Quinn and Kevin Smith (2007). 'Cenozoic pull-apart basins in the British Isles', British Geological Survey, Bicentennial Poster, http://nora.nerc.ac.uk/id/eprint/517298/1/BicentennialPoster_PullApartBasins.pdf

4. Nick M.W. Roberts, Jack K. Lee, Robert E. Holdsworth, Christopher Jeans, Andrew R. Farrant, and Richard Haslam (2020). 'Near-surface Palaeocene fluid flow, mineralisation and faulting at Flamborough Head, UK: new field observations and U-Pb calcite dating constraints.' *Solid Earth Discussions*, 11 (5), 1931–1945, https://www.solid-earth-discuss.net/se-2020-73/se-2020-73.pdf
5. 'Tertiary rivers: Oligocene', *Cambridge Palaeoenvironments Group*, https://www.qpg.geog.cam.ac.uk/research/projects/tertiaryrivers/oligoc.html
6. J.P. Lefort and B.N.P. Agarwal (2000). 'Gravity and geomorphological evidence for a large crustal bulge cutting across Brittany (France): a tectonic response to the closure of the Bay of Biscay.' *Tectonophysics*, 323 (3–4), 149–162, https://www.sciencedirect.com/science/article/pii/S0040195100001037
7. J.L. Jeager (1967). 'Un alignement d'anomalies légères coincidant avec des bassins tertiaires en Bretagne.' *Mémoires du Bureau de recherches géologiques et minières*, 52, 89–102.
8. M.C. Boulter and D.L. Craig (1979). 'A Middle Oligocene pollen and spore assemblage from the Bristol Channel.' *Review of Palaeobotany and Palynology*, 28 (3–4), 259–272, https://doi.org/10.1016/0034-6667(79)90028-9
9. D. Graham Jenkins, M.C. Boulter, and A.T.S. Ramsay (1995). 'The Flimston Clay, Pembrokeshire, Wales: a probable late Oligocene lacustrine deposit.' *Journal of Micropalaeontology*, 14 (1), 66, http://jm.lyellcollection.org/content/14/1/66.abstract
10. M.R. Dobson, L. Delanty, and R.J. Whittington (1982). 'Stratigraphy and inversion tectonics in the St George's Channel area off SW Wales, UK.' *Geo-Marine Letters*, 2 (1–2), 23–30, https://doi.org/10.1007/BF02462796; Simon P. Holford, Paul F. Green, Jonathan P. Turner, Gareth A. Williams, David R. Tappin, and Ian R. Duddy (2008). 'Evidence for km-scale Neogene exhumation driven by compressional deformation in the Irish Sea basin system.' In: H. Johnson, A.G. Doré, R.W. Gatliff, R. Holdsworth, E.R. Lundin, and J.D. Ritchie (eds), *The Nature and Origin of Compression in Passive Margins*, London: The Geological Society, Special Publications no. 306, pp. 91–119, https://core.ac.uk/download/pdf/62250.pdf
11. John Andrew Fitzgerald (1999). 'Pollen and spore assemblages from the Oligocene Lough Neagh Group and Dunaghy Formation, Northern Ireland.' PhD thesis, Centre for Palynology, University of Sheffield, http://etheses.whiterose.ac.uk/10365/1/523165_vol1.pdf
12. John Parnell, Balvinder Shukla, and Ian G. Meighan (1989). 'The lignite and associated sediments of the Tertiary Lough Neagh Basin.' *Irish Journal of Earth Sciences*, 10 (1), 67–88, https://www.jstor.org/stable/30002250

13. Parnell et al., 'Lignite and associated sediments'.
14. G.C. Wilkinson, R.A.B. Bazley, and M.C. Boulter (1980). 'The geology and palynology of the Oligocene Lough Neagh Clays, Northern Ireland.' *Journal of the Geological Society*, 137 (1), 65–75, https://doi.org/10.1144/gsjgs.137.1.0065
15. W.I. Mitchell (ed.) (2004). 'Lough Neagh Group, Late Palaeogene (Oligocene), Northern Ireland.' In: *The Geology of Northern Ireland: Our Natural Foundation*, Belfast: Geological Survey of Northern Ireland, http://earthwise.bgs.ac.uk/index.php/Lough_Neagh_Group,_Late_Palaeogene_(Oligocene),_Northern_Ireland
16. David K. Smythe and Nicholas Kenolty (1975). 'Tertiary sediments in the Sea of the Hebrides.' *Journal of the Geological Society*, 131 (2), 227–233, https://doi.org/10.1144/gsjgs.131.2.0227
17. R.W. England (1994). 'The structure of the Skye Lava field.' *Scottish Journal of Geology*, 30 (1), 33–37, https://doi.org/10.1144/sjg30010033
18. D. Evans, C. Hallsworthy, D.W. Jolley, and A.C. Morton (1991). 'Late Oligocene terrestrial sediments from a small basin in the Little Minch.' *Scottish Journal of Geology*, 27, 33–40, https://doi.org/10.1144/sjg27010033
19. D. Evans, A.C. Morton, S. Wilson, and B.A. Barreiro (1997). 'Palaeoenvironmental significance of marine and terrestrial Tertiary sediments on the NW Scottish shelf in BGS borehole 77/7.' *Scottish Journal of Geology*, 33 (1), 31–42, https://doi.org/10.1144/sjg33010031; D. Evans, J.A. Chesher, C.E. Deegan, and N.G.T. Fannin (1981). 'The offshore geology of Scotland in relation to the IGS shallow drilling programme, 1970–1978', Report of the Institute of Geological Sciences No. 81/12.
20. T. Neville George (1966). 'Geomorphic evolution in Hebridean Scotland.' *Scottish Journal of Geology*, 2 (1), 1–34, https://doi.org/10.1144/sjg02010001
21. M.R. Cooper, H. Anderson, J.J. Walsh, C.L. Van Dam, M.E. Young, G. Earls, and A. Walker (2012). 'Palaeogene Alpine tectonics and Icelandic plume-related magmatism and deformation in Northern Ireland.' *Journal of the Geological Society*, 169 (1), 29–36, https://doi.org/10.1144/0016-76492010-182
22. Parnell et al., 'Lignite and associated sediments'.
23. Lars Ole Boldreel and Morten Sparre Andersen (1994). 'Tertiary development of the Faeroe-Rockall Plateau based on reflection seismic data.' *Bulletin of the Geological Society of Denmark*, 41, 162–180; J. Derek Ritchie, Howard Johnson, Martyn F. Quinn, and Robert W. Gatliff (2008). 'The effects of Cenozoic compression within the Faroe–Shetland Basin and adjacent areas.' In: H. Johnson, A.G. Doré, R.W. Gatliff, R. Holdsworth, E.R. Lundin, and J.D. Ritchie (eds), *The Nature and Origin of Compression*

in Passive Margins, London: The Geological Society, Special Publications no. 306, pp. 121–136; H. Johnson, J.D. Ritchie, K. Hitchen, D.B. McInroy, and G.S. Kimbell (2005). 'Aspects of the Cenozoic deformational history of the Northeast Faroe–Shetland Basin, Wyville–Thomson Ridge and Hatton Bank areas.' *Geological Society, London, Petroleum Geology Conference series*, 6, 993–1007, https://doi.org/10.1144/0060993; J.D. Ritchie, H. Johnson, and G.S. Kimbell (2003). 'The nature and age of Cenozoic contractional deformation within the NE Faroe–Shetland Basin.' *Marine and Petroleum Geology*, 20 (5), 399–409, https://www.sciencedirect.com/science/article/pii/S0264817203000758

24. David Ellis and Martyn S. Stoker (2014). 'The Faroe–Shetland Basin: a regional perspective from the Palaeocene to the present day and its relationship to the opening of the North Atlantic Ocean.' In: S.J.C. Cannon and D. Ellis (eds), *Hydrocarbon Exploration to Exploitation West of Shetlands*, London: The Geological Society, Special Publications no. 397, pp. 11–31, http://nora.nerc.ac.uk/id/eprint/503195/1/Ellis%20and%20Stoker_SP397_2014.pdf

CHAPTER 8

1. P. Walsh, I. Morawiecka, and Krystyna Skawińska-Wieser (1996). 'A Miocene palynoflora preserved by karstic subsidence in Anglesey and the origin of the Menaian Surface.' *Geological Magazine*, 133 (6), 713–719, https://doi.org/10.1017/S0016756800024560
2. P.T. Walsh, M.C. Boulter, M. Ijtaba, and D.M. Urbani (1972). 'The preservation of the Neogene Brassington Formation of the southern Pennines and its bearing on the evolution of Upland Britain.' *Journal of the Geological Society*, 128 (6), 519–559, https://doi.org/10.1144/gsjgs.128.6.0519
3. M. Pound and J. Riding (2015). 'Miocene in the UK!', *Geoscientist Online*, https://www.geolsoc.org.uk/Geoscientist/Archive/March-2015/Miocene-in-the-UK
4. P.T. Walsh, K. Atkinson, M.C. Boulter, and R.A. Shakesby (1987). 'The Oligocene and Miocene outliers of west Cornwall and their bearing on the geomorphological evolution of Oldland Britain.' *Philosophical Transactions of the Royal Society A*, 323 (1571), 211–245.
5. Bernard O'Connor (2009). *The Suffolk Fossil Diggings*, pp. 55–56.
6. Stephen Louwye, Robert Marquet, Mark Bosselaers, and Olivier Lambert (2010). 'Stratigraphy of an Early–Middle Miocene sequence near Antwerp in northern Belgium (southern North Sea Basin).' *Geologica Belgica*, 13 (3), 269–284.

7. O'Connor, *The Suffolk Fossil Diggings*, p. 49.
8. O'Connor, *The Suffolk Fossil Diggings*, p. 223.
9. 'GeoSuffolk: the Deben Estuary', Suffolk RIGS Group, https://geosuffolk.co.uk/images/SuffolkGeoCoast/DebenLeaflet.pdf
10. Tim Holt-Wilson (2014). 'Boxstone', *Our Vital Earth Blog*, https://storvaxt.blogspot.co.uk/2014/05/
11. D.J.J. van Hinsbergen, R.L.M. Vissers, and W. Spakman (2014). 'Origin and consequences of western Mediterranean subduction, rollback, and slab segmentation,' *Tectonics*, 33 (4), 393–419, http://onlinelibrary.wiley.com/doi/10.1002/2013TC003349/pdf
12. Pierre-Charles de Granciansky, David G. Roberts, and Pierre Tricart (2011). *The Western Alps from Rift to Passive Margin to Orogenic Belt: An Integrated Geoscience Overview*. Oxford: Elsevier.
13. J.F. Dewey (2000). 'Cenozoic tectonics of western Ireland.' *Proceedings of the Geologists' Association*, 111 (4), 291–306, https://doi.org/10.1016/S0016-7878(00)80086-3
14. V. Rime, A. Sommaruga, M. Schori, *et al.* (2019). 'Tectonics of the Neuchâtel Jura Mountains: insights from mapping and forward modelling.' *Swiss Journal of Geosciences* 112, 563–578, https://doi.org/10.1007/s00015-019-00349-y; Herfried Madritsch (2008). 'Structural evolution and neotectonics of the Rhine-Bresse transfer zone', PhD dissertation, University of Basel, http://edoc.unibas.ch/856/1/DissB_8470.pdf; Stéphane Molliex, Olivier Fabbri, Vincent Bichet, and Herfried Madritsch (2011). 'Possible Quaternary growth of a hidden anticline at the front of the Jura fold-and-thrust belt: geomorphological constraints from the Forêt de Chaux area, France.' *Bulletin de la Société Géologique de France*, 182 (4), 337–346, https://doi.org/10.2113/gssgfbull.182.4.337, http://hal.ird.fr/file/index/docid/633369/filename/Molliex_et_al_BSGF_final-1.pdf
15. Muriel Rocher, Marc Cushing, Francis Lemeille, and Stéphane Baize (2005). 'Stress induced by the Mio-Pliocene Alpine collision in northern France.' *Bulletin de la Société Géologique de France*, 176 (4), 319–328, https://doi.org/10.2113/176.4.319
16. Y. Rotstein and M. Schaming (2011). 'The Upper Rhine Graben (URG) revisited: Miocene transtension and transpression account for the observed first-order structures.' *Tectonics*, 30 (3), https://doi.org/10.1029/2010TC002767
17. David Ellis and Martyn S. Stoker (2014). 'The Faroe–Shetland Basin: a regional perspective from the Paleocene to the present day and its relationship to the opening of the North Atlantic Ocean.' In: S.J.C. Cannon and D. Ellis (eds), *Hydrocarbon Exploration to Exploitation West of Shetlands*,

London: The Geological Society, Special Publications no. 397, pp. 11–31, https://doi.org/10.1144/SP397.1

18. J. Derek Ritchie, Howard Johnson, Martyn F. Quinn, and Robert W. Gatliff (2008). 'The effects of Cenozoic compression within the Faroe-Shetland Basin and adjacent areas.' In: H. Johnson, A.G. Doré, R.W. Gatliff, R. Holdsworth, E.R. Lundin, and J.D. Ritchie (eds), *The Nature and Origin of Compression in Passive Margins*, London: The Geological Society, Special Publications no. 306, pp. 121–136, http://sp.lyellcollection.org/content/306/1/121; Aage Bach Sørensen (2003). 'Cenozoic basin development and stratigraphy of the Faroes area.' *Petroleum Geoscience*, 9 (3), 189–207, https://doi.org/10.1144/1354-079302-508; L.O. Boldreel and M.S. Andersen (1993). 'Late Paleocene to Miocene compression in the Faeroe-Rockall area.' In: J.R. Parker (ed.), *Petroleum Geology of North West Europe: Proceedings of the 4th Conference*, London: The Geological Society, pp. 1025–1034.
19. Morten Sparre Andersen, Tove Nielsen, Aage Bach Sørensen, Lars Ole Boldreel, and Anton Kuijpers (2000). 'Cenozoic sediment distribution and tectonic movements in the Faroe region.' *Global and Planetary Change*, 24 (3–4), 239–259, https://www.sciencedirect.com/science/article/pii/S0921818100000114; Jana Ólavsdóttir, Óluva R. Eidesgaard, and Martyn S. Stoker (2016). 'The stratigraphy and structure of the Faroese continental margin.' In: G. Péron-Pinvidic, J.R. Hopper, M.S. Stoker, C. Gaina, J.C. Doornenbal, T. Funck, and U.E. Árting (eds), *The NE Atlantic Region: A Reappraisal of Crustal Structure, Tectonostratigraphy and Magmatic Evolution*, London: The Geological Society, Special Publications no. 447, pp. 339–356, https://sp.lyellcollection.org/content/447/1/339
20. Nick M.W. Roberts and Richard J. Walker (2016). 'U-Pb geochronology of calcite-mineralized faults: absolute timing of rift-related fault events on the NE Atlantic margin.' *Geology*, 44 (7), 531–534, https://doi.org/10.1130/G37868.1
21. Ritchie et al., 'Effects of Cenozoic compression'.
22. H. Løseth, B. Raulline, and A. Nygård (2013). 'Late Cenozoic geological evolution of the northern North Sea: development of a Miocene unconformity reshaped by large-scale Pleistocene sand intrusion.' *Journal of the Geological Society*, 170 (1), 133–145, https://doi.org/10.1144/jgs2011-165
23. H. Løseth and S. Henriksen (2005). 'A Middle to Late Miocene compression phase along the Norwegian passive margin.' *Geological Society, London, Petroleum Geology Conference series*, 6, 845–859, https://doi.org/10.1144/0060845; P.A. Nadin, N.J. Kusznir, and M.J. Cheadle (1997). 'Early Tertiary plume uplift of the North Sea and Faeroe-Shetland

Basins.' *Earth and Planetary Science Letters*, 148 (1–2), 109–127, https://www.sciencedirect.com/science/article/pii/S0012821X97000356

24. Yngve Rundberg and Tor Eidvin (2005). 'Controls on depositional history and architecture of the Oligocene–Miocene succession, northern North Sea Basin.' In: B.T.G. Wandas, J.P. Nystuen, E.A. Eide, and F.M. Gradstein (eds), *Onshore–Offshore Relationships on the North Atlantic Margin*, Norwegian Petroleum Society Special Publication 12, Amsterdam: Elsevier, pp. 207–239, http://www.npd.no/Global/Norsk/3-Publikasjoner/Forskningsartikler/Rundberg-Eidvin-2005.pdf
25. David B. Jordan and Ray Burgess (2007). 'A Miocene fault in south-east Ireland revealed by ^{40}Ar-^{39}Ar dating of hydrothermal cryptomelane.' *Irish Journal of Earth Sciences*, 25, 55–61.
26. D.K. Smythe (1987). 'Seismic studies in the region of northern Skye and the Little Minch', PhD thesis, University of Glasgow, pp. 57 and 69.
27. D.G. Roberts (1989). 'Basin inversion in and around the British Isles.' In: M.A. Cooper and D.G. Williams (eds), *Inversion Tectonics*, London: The Geological Society, Special Publications no. 44, pp. 131–150.
28. R.W. England (1994). 'The structure of the Skye Lava field.' *Scottish Journal of Geology*, 30 (1), 33–37, https://doi.org/10.1144/sjg30010033
29. Nicol Morton and J.D. Hudson (1995). 'Field guide to the Jurassic of the Isles of Raasay and Skye, Inner Hebrides, NW Scotland.' In: J.D. Taylor (ed.), *Field Geology of the British Jurassic*, London: The Geological Society, pp. 209–280.
30. Dave Petley (2012). 'The most beautiful landslide on Earth? The Quiraing landslide on the Isle of Skye, Scotland', *The Landslide Blog*, https://blogs.agu.org/landslideblog/2012/11/06/the-most-beautiful-landslide-on-earth-the-quiraing-landslide-on-the-isle-of-skye-scotland/
31. K.C. Anderson and F.W. Dunham (1966). *The Geology of Northern Skye, Memoirs of the Geological Survey of Great Britain (Scotland) (District)*. Edinburgh: Institute of Geological Sciences, HMSO, p. 174.
32. Roberts, 'Basin inversion'.
33. R. Shelton (1997). 'Tectonic evolution of the Larne Basin.' In: N.S. Meadows, S.P. Trueblood, M. Hardman, and G. Cowan (eds), *Petroleum Geology of the Irish Sea and Adjacent Areas*, London: The Geological Society, Special Publications no. 124, pp. 113–133.
34. Simon P. Holford, Paul F. Green, Richard R. Hillis, Jonathan P. Turner, and Carl T.E. Stevenson (2009). 'Mesozoic–Cenozoic exhumation and volcanism in Northern Ireland constrained by AFTA and compaction data from the Larne No. 2 borehole.' *Petroleum Geoscience*, 15 (3), 239–257, https://doi.org/10.1144/1354-079309-840

35. Gareth A. Williams, Jonathan P. Turner, and Simon P. Holford (2005). 'Inversion and exhumation of the St. George's Channel Basin, offshore Wales, UK.' *Journal of the Geological Society*, 162 (1), 97–110, https://doi.org/10.1144/0016-764904-023, http://nora.nerc.ac.uk/id/eprint/18234/1/SGCB.pdf
36. Richard R. Hillis, Simon P. Holford, Paul F. Green, Anthony G. Doré, Robert W. Gatliff, Martyn S. Stoker, Kenneth Thomson, Jonathan P. Turner, John R. Underhill, and Gareth A. Williams (2008). 'Cenozoic exhumation of the southern British Isles.' *Geology*, 36 (5), 371–374, https://doi.org/10.1130/G24699A.1
37. S.P. Holford, P.F. Green, J.P. Turner, G.A. Williams, D.R. Tappin, and I.R. Duddy (2008). 'Evidence for km-scale Neogene exhumation driven by compressional deformation in the Irish Sea basin system.' In: Howard Johnson, Tony G. Doré, Robert W. Gatliff, Robert W. Holdsworth, Erik R. Lundin, and J. Derek Ritchie (eds), *The Nature and Origin of Compression in Passive Margins*, London: The Geological Society, Special Publications no. 306, pp. 91–119, http://nora.nerc.ac.uk/id/eprint/4912/1/Holford_et_al_COMPRESSIONAL_MARGINS_full_paper.pdf
38. S.P. Holford, P.F. Green, and J.P. Turner (2005). 'Palaeothermal and compaction studies in the Mochras borehole (NW Wales) reveal early Cretaceous and Neogene exhumation and argue against regional Palaeogene uplift in the southern Irish Sea.' *Journal of the Geological Society*, 162, 829–840.
39. P. Walsh, M. Boulter, and I. Morawiecka (1999). 'Chattian and Miocene elements in the modern landscape of western Britain and Ireland.' In: B.J. Smith, W.B. Whalley, and P.A. Warke (eds), *Uplift, Erosion and Stability: Perspectives on Long-Term Landscape Development*, London: The Geological Society, Special Publications no. 162, pp. 45–63.
40. Michael J.M. Cunningham, Adrian W.E. Phillips, and Alexander L. Densmore (2004). 'Evidence for Cenozoic tectonic deformation in SE Ireland and near offshore.' *Tectonics*, 23 (6), TC6002, https://doi.org/10.1029/2003TC001597
41. L.M. Murdoch, F.W. Musgrove, and J.S. Perry (1995). 'Tertiary uplift and inversion history in the North Celtic Sea Basin and its influence on source rock maturity.' In: P.F. Croker and P.M. Shannon (eds), *The Petroleum Geology of Ireland's Offshore Basins*, London: The Geological Society, Special Publications no. 93, pp. 297–319, http://sp.lyellcollection.org/content/93/1/297.abstract; Roger M. Tucker and Graham Arter (1987). 'The tectonic evolution of the North Celtic Sea and

Cardigan Bay basins with special reference to basin inversion.' *Tectonophysics*, 137 (1–4), 291–307, http://www.sciencedirect.com/science/article/pii/0040195187903246

42. Edward Greenly (1937). 'The age of the mountains of Snowdonia.' In: *The Geology of Snowdonia, a Collection of Articles on the Physical Features of the Peaks of Snowdonia*, Read Books Ltd, 2011.
43. R.E. Bevins, J.M. Horak, A.D. Evans, and R. Morgan (1996). 'Palaeogene dyke swarm, NW Wales: evidence for Cenozoic sinistral fault movement.' *Journal of the Geological Society*, 153 (2), 177–180, https://doi.org/10.1144/gsjgs.153.2.0177
44. R.A. Chadwick, D.J. Evans, and D.W. Holliday (1993). 'The Maryport fault: the post-Caledonian tectonic history of southern Britain in microcosm.' *Journal of the Geological Society*, 150 (2), 247–250, https://doi.org/10.1144/gsjgs.150.2.0247
45. P. Stone, D. Millward, B. Young, J.W. Merritt, S.M. Clarke, M. McCormac, and D.J.D. Lawrence (2010). 'Late Mesozoic and Cenozoic tectonics and magmatism, northern England.' In: *British Regional Geology: Northern England*, fifth edition, Keyworth, Nottingham: British Geological Survey, http://earthwise.bgs.ac.uk/index.php/Late_Mesozoic_and_Cenozoic_tectonics_and_magmatism,_Northern_England
46. P.J. Newman (1999). 'The geology and hydrocarbon potential of the Peel and Solway Basins, East Irish Sea.' *Journal of Petroleum Geology*, 22 (3), 305–324.
47. T.C. Pharaoh, C.M.A. Gent, S.D. Hannis, K.L. Kirk, A.A. Monaghan, M.F. Quinn, N.J.P. Smith, C.H. Vane, O. Wakefield, and C.N. Waters (2018). 'An overlooked play? Structure, stratigraphy and hydrocarbon prospectivity of the Carboniferous in the East Irish Sea-North Channel basin complex.' In: A.A. Monaghan, J.R. Underhill, A.J. Hewett, and J.E.A. Marshall (eds), *Paleozoic Plays of NW Europe*, London: The Geological Society, Special Publications no. 471, pp. 281–316, http://nora.nerc.ac.uk/id/eprint/520263/1/GSLSpecPub17-140_R2.pdf
48. G.I. Lumsden and A. Davies (1965). 'The buried channel of the River Nith and its marked change in level across the Southern Upland Fault.' *Scottish Journal of Geology*, 1 (2), 134–143, https://doi.org/10.1144/sjg01020134
49. A. Vincent (1974). 'Sedimentary environments of the Bovey Basin', MPhil thesis, University of Surrey; A. Vincent (1983). 'The origin and occurrence of Devon Ball Clays.' In: R.C.L. Wilson (ed.), *Residual Deposits: Surface Related Weathering Processes and Materials*, London: The Geological Society, Special Publications no. 11, pp. 39–45, https://doi.org/10.1144/GSL.SP.1983.011.01.06

50. S. Holloway and R.A. Chadwick (1986). 'The Sticklepath-Lustleigh fault zone: Tertiary sinistral reactivation of a Variscan dextral strike-slip fault.' *Journal of the Geological Society*, 143 (3), 447–452, https://jgs.lyellcollection.org/content/143/3/447
51. Colin M. Bristow and D.E. Hughes (1971). 'A Tertiary thrust fault on the southern margin of the Bovey Basin.' *Geological Magazine*, 108 (1), 61–67, https://doi.org/10.1017/S0016756800050974
52. C.M. Bristow, Q.G. Palmer, D. Pirrie Bristow, and Duncan Pirrie (1992). 'Palaeogene basin development: new evidence from the southern Petrockstow Basin, Devon.' *Geoscience in South-West England*, 8 (1), 18–22.
53. J.B. Butler (1990). 'A review of the tectonic history of the Shropshire area.' *Proceedings of the Shropshire Geological Society*, 9, 20–34, http://www.shropshiregeology.org.uk/sgspublications/Proceedings/1990%20No_09%20020-034%20Butler%20tectonic%20history.pdf; A.M. Hall (1991). 'Pre-Quaternary landscape evolution in the Scottish Highlands.' *Earth and Environmental Science Transactions of the Royal Society of Edinburgh*, 82 (1), 1–26, https://doi.org/10.1017/S0263593300007495
54. W.H. Zagwijn (1989). 'The Netherlands during the Tertiary and the Quaternary: a case history of Coastal Lowland evolution.' *Geologie en Mijnbouw*, 68, 107–120, https://doi.org/10.1007/978-94-017-1064-0_6
55. T.G. Bevan and P.L. Hancock (1986). 'A late Cenozoic regional mesofracture system in southern England and northern France.' *Journal of the Geological Society*, 143 (2), 355–362, https://jgs.lyellcollection.org/content/143/2/355
56. A.F. Howland (1991). 'London's Docklands: engineering geology.' *Proceedings of the Institution of Civil Engineers, Part 1*, 90 (6), 1153–1178; Katherine Royse (2009). 'New insights into the geology under London through the analysis of 3D models.' In: *EUREGEO 2009: European Congress on Regional Geoscientific Cartography and Information Systems, Munich, Germany, 9–12 June 2009*, Munich, Germany: Bayerisches Landesamt für Umwelt Referat GeoForum Bayern, Geotopkataster, 1–4, https://nora.nerc.ac.uk/id/eprint/7613/
57. V.J. Banks, S.H. Bricker, K.R. Royse, and P.E.F. Collins (2015). 'Anomalous buried hollows in London: development of a hazard susceptibility.' *Quarterly Journal of Engineering Geology and Hydrogeology*, 48, 55–70, https://doi.org/10.1144/qjegh2014-037, http://bura.brunel.ac.uk/bitstream/2438/11469/3/Fulltext.pdf
58. Henry Dewey and C.E.N. Bromehead (1915). *The Geology of the Country around Windsor and Chertsey.* Memoirs of the Geological Survey, Sheet 269.

59. Cristina Persano and Katherine J. Dobson (2009). 'How many years can a mountain exist before it's washed to the sea? Or, Bob Dylan's theory of landscape evolution.' *Scottish Geographical Journal*, 125 (3–4), 370–378, https://doi.org/10.1080/14702540903364419
60. Martyn S. Stoker, Simon P. Holford, Richard R. Hillis, Paul F. Green, and Ian R. Duddy (2010). 'Cenozoic post-rift sedimentation off northwest Britain: recording the detritus of episodic uplift on a passive continental margin.' *Geology*, 38 (7), 595–598, https://core.ac.uk/download/pdf/57401.pdf

CHAPTER 9

1. Martin Ekman (2009). 'The changing level of the Baltic Sea during 300 years: a clue to understanding the Earth', Summer Institute for Historical Geophysics, Åland Islands, http://www.historicalgeophysics.ax/The%20Changing%20Level%20of%20the%20Baltic%20Sea.pdf
2. Charles Lyell (1835). 'On the proofs of a gradual rising of the land in certain parts of Sweden, The Bakerian Lecture.' *Philosophical Transactions of the Royal Society of London*, 125, 1–38, https://royalsocietypublishing.org/doi/10.1098/rstl.1835.0002
3. Roy Shepherd and Robert Randall (2002). 'Walton on the Naze (Essex)', *Discovering Fossils*, http://www.discoveringfossils.co.uk/walton_on_naze_fossils.htm
4. P.J. Boylan (1998). 'Lyell and the dilemma of Quaternary glaciation.' In: D.J. Blundell and A.C. Scott (eds), *Lyell: The Past is the Key to the Present*, London: The Geological Society, Special Publications no. 143, pp. 145–159, http://sp.lyellcollection.org/content/specpubgsl/143/1/145.full.pdf
5. T. Jamieson (1865). 'On the history of the last geological changes in Scotland.' *Quarterly Journal of the Geological Society of London*, 21 (1–2), 161–203, https://doi.org/10.1144/GSL.JGS.1865.021.01-02.24
6. In a letter to Lyell on 14 September 1838.
7. Diary entry 22 September 1773, James Boswell, *The Journal of a Tour to the Hebrides with Samuel Johnson, LL.D.*
8. M. Böse, C. Lüthgens, J.R. Lee, and J. Rose (2012). 'Quaternary glaciations of northern Europe.' *Quaternary Science Reviews* (44), 17. CiteSeerX 10.1.1.734.1691
9. V.J. Banks, S.H. Bricker, K.R. Royse, and P.E.F. Collins (2015). 'Anomalous buried hollows in London: development of a hazard susceptibility map.' *Quarterly Journal of Engineering Geology and Hydrogeology*, 48 (1), 55–70, https://qjegh.lyellcollection.org/content/48/1/55

10. David J.A. Evans, Mark D. Bateman, David H. Roberts, Alicia Medialdea, Laura Hayes, Geoff A.T. Duller, Derek Fabel, and Chris D. Clark (2016). 'Glacial Lake Pickering: stratigraphy and chronology of a proglacial lake dammed by the North Sea Lobe of the British–Irish Ice Sheet.' *Journal of Quaternary Science*, 32 (2), 295–310, https://doi.org/10.1002/jqs.2833
11. A. Straw (2016). 'Devensian glaciers and lakes in Lincolnshire and S. Yorkshire.' *Mercian Geologist*, 19 (1), 39–46, https://www.thecollectionmuseum.com/assets/downloads/Straw_Devensian_glaciers_and_lakes_in_Lincs_and_Yorks.pdf
12. Della K. Murton and Julian B. Murton (2012). 'Middle and Late Pleistocene glacial lakes of lowland Britain and the southern North Sea Basin.' *Quaternary International*, 260, 115–142, https://doi.org/10.1016/j.quaint.2011.07.034
13. G.W. Green (1992). 'Quaternary events in the Lower Severn and Avon valleys.' In: *British Regional Geology: Bristol and Gloucester Region*, third edition, London: HMSO for the British Geological Survey, http://earthwise.bgs.ac.uk/index.php/Quaternary_events_in_the_Lower_Severn_and_Avon_valleys
14. C.P. Green, D.F.M. McGregor, and A.H. Evans (1982). 'Development of the Thames drainage system in Early and Middle Pleistocene times.' *Geological Magazine*, 119 (3), 281–290, https://doi.org/10.1017/S0016756800026091
15. Sanjeev Gupta, Jenny S. Collier, David Garcia-Moreno, Francesca Oggioni, Alain Trentesaux, Kris Vanneste, Marc De Batist, Thierry Camelbeeck, Graeme Potter, Brigitte Van Vliet-Lanoë, and John C.R. Arthur (2017). 'Two-stage opening of the Dover Strait and the origin of island Britain.' *Nature Communications*, 8: art. no. 15101, https://doi.org/10.1038/ncomms15101
16. Alec J. Smith (1985). 'A catastrophic origin for the palaeovalley system of the eastern English Channel.' *Marine Geology*, 64 (1–2), 65–75, https://doi.org/10.1016/0025-3227(85)90160-4
17. P. Gibbard (2007). 'Europe cut adrift.' *Nature*, 448 (7151), 259–260, https://doi.org/10.1038/448259a; P. Gibbard (2009). 'How Britain became an island: the report', *Cambridge Quaternary Paleoenvironments Group*, https://www.qpg.geog.cam.ac.uk/research/projects/englishchannelformation/
18. D. Hamilton and A.J. Smith (1971). 'The origin and sedimentary history of the Hurd Deep, English Channel, with additional notes on other deeps in the western English Channel.' *Mémoires du Bureau de recherches géologiques et minières*, 79, 59–83.

19. Berit Oline Hjelstuen, Hans Petter Sejrup, Espen Valvik, and Lukas W.M. Becker (2018). 'Evidence of an ice-dammed lake outburst in the North Sea during the last deglaciation.' *Marine Geology*, 402, 118–130, https://www.sciencedirect.com/science/article/abs/pii/S0025322717301391
20. K. Lambeck (1996). 'Glaciation and sea-level change for Ireland and the Irish Sea since Late Devensian/Midlandian time.' *Journal of the Geological Society*, 153 (6), 853–872, https://doi.org/10.1144/gsjgs.153.6.0853
21. J.M. Gray (1978). 'Low-level shore platforms in the south-west Scottish Highlands: altitude, age and correlation.' *Transactions of the Institute of British Geographers*, 3 (2), 151–164.
22. R. Muir-Wood (1989). 'Extraordinary deglaciation reverse faulting in northern Fennoscandia.' In: S. Gregersen and P.W. Basham (eds), *Earthquakes at North Atlantic Passive Margins: Neotectonics and Post-Glacial Rebound*, Dordrecht: Kluwer, pp. 141–174.
23. R. Muir-Wood (2000). 'Deglaciation seismotectonics: a principal influence on intraplate seismogenesis at high latitudes.' *Quaternary Science Reviews*, 19 (14–15), 1399–1411.

CHAPTER 10

1. John Betjeman (1952). *First and Last Loves*, London: Faber & Faber.
2. Letter from Charles Robert Darwin to Horace Darwin, 26 July 1868, Darwin Correspondence Project, https://www.darwinproject.ac.uk/letter/DCP-LETT-6289.xml
3. R. Holmes, P.R.N. Hobbs, A.B. Leslie, I.P. Wilkinson, F.J. Gregory, J.B. Riding, R.J. Hoult, R.M. Cooper, and S.M. Jones (2003). 'DTI Strategic Environmental Assessment Area 4 (SEA4): Geological Evolution Pilot Whale Diapirs and Stability of the Seabed Habitat', British Geological Survey Commercial Report CR/03/082, http://nora.nerc.ac.uk/id/eprint/509851/1/CR03082N.pdf
4. L. Ahorner (1962). 'Untersuchungen zur Quartären Bruchtektonik der Niederrheinischen Bucht: Eiszeitalter und Gegenwart.' *Quaternary Science Journal*, 13 (1), 24–105, https://doi.org/10.23689/fidgeo-1466; K.-G. Hinzen and S.K. Reamer (2007). 'Seismicity, seismotectonics, and seismic hazard in the northern Rhine area.' In: S. Stein and S. Mazotti (eds), *Continental Intraplate Earthquakes: Science, Hazard, and Policy Issues*, Boulder, CO: Geological Society of America, Special Paper no. 425, pp. 225–242, https://doi.org/10.1130/2007.2425(15)

5. Zoe K. Shipton, Mustapha Meghraoui, and Louise Monro (2016). 'Seismic slip on the west flank of the Upper Rhine Graben (France–Germany): evidence from tectonic morphology and cataclastic deformation bands.' In: A. Landgraf, S. Kübler, E. Hintersberger, and S. Stein (eds), *Seismicity, Fault Rupture and Earthquake Hazards in Slowly Deforming Regions*, London: The Geological Society, Special Publications no. 432, pp. 147–161, http://sp.lyellcollection.org/content/432/1/147
6. Peter Japsen and James A. Chalmers (2000). 'Neogene uplift and tectonics around the North Atlantic: overview.' *Global and Planetary Change*, 24 (3–4), 165–173, https://doi.org/10.1016/S0921-8181(00)00006-0
7. A. Dalland (1975). 'The Mesozoic rocks of Andøy, northern Norway.' *Norges geologiske Undersøkelse*, 316, 271–287.
8. C.K. Brooks (1985). 'Vertical crustal movements in the Tertiary of central East Greenland: a continental margin at a hot-spot.' *Zeitschrift für Geomorphologie*, 54, 101–117.
9. Gilbert White (1789). *Natural History and Antiquities of Selborne.*
10. I. Overeem, G.J. Weltje, C. Bishop-Kay, and S.B. Kroonenberg (2001). 'The Late Cenozoic Eridanos delta system in the southern North Sea Basin: a climate signal in sediment supply?' *Basin Research*, 13 (3), 293–312, https://doi.org/10.1046/j.1365-2117.2001.00151.x
11. Peter Frank Riches (2012). 'The palaeoenvironmental and neotectonic history of the Early Pleistocene Crag basin in East Anglia', PhD thesis, Royal Holloway University of London.
12. H.G. Dines and C.P. Chatwin (1930). 'Pliocene sandstone from Rothamsted (Hertfordshire).' In: *Geological Survey Summary of Progress for 1929*, pp. 1–7.
13. John Catt (2012). 'New excavation at Little Heath geological SSSI.' *Transactions of the Hertfordshire Natural History Society*, 44 (1), 68, https://ptes.org/wp-content/uploads/2016/07/Lichens-in-Herts-OrchardsHerts-Naturalist-20121.pdf; John Catt (2013). 'Dating marine gravels in Hertfordshire: John Catt leads Quaternary Research Association field meeting', *UCL Department of Geography*, https://www.geog.ucl.ac.uk/news-events/news/news-archive/2013/may-2013/dating-marine-gravels-in-hertfordshire
14. A.J. Moffatt, J.A. Catt, R. Webster, and E.H. Brown (1986). 'A re-examination of the evidence for a Plio-Pleistocene marine transgression on the Chiltern Hills. III. Deposits.' *Earth Surface Processes and Landforms*, 11 (3), 233–247, https://doi.org/10.1002/esp.3290110302
15. R.C. Preece, J.D. Scourse, S.D. Houghton, K.L. Knudsen, and D.N. Penney (1990). 'The Pleistocene sea-level and neotectonic history

of the eastern Solent, southern England.' *Philosophical Transactions of the Royal Society B: Biological Sciences*, 328 (1249), 425–477, https://doi.org/10.1098/rstb.1990.0120

16. Pierre Antoine, Jean Pierre Lautridou, and Michel Laurent (2000). 'Long-term fluvial archives in NW France: response of the Seine and Somme rivers to tectonic movements, climatic variations and sea-level changes.' *Geomorphology*, 33 (3–4), 183–207, https://doi.org/10.1016/S0169-555X(99)00122-1
17. A.B. Watts, W.S. McKerrow, and E. Fielding (2000). 'Lithospheric flexure, uplift, and landscape evolution in south-central England.' *Journal of the Geological Society*, 157 (6), 1169–1177, https://doi.org/10.1144/jgs.157.6.1169; N.F. Lane, A.B. Watts, and Andrew R. Farrant (2008). 'An analysis of Cotswold topography: insights into the landscape response to denudational isostasy.' *Journal of the Geological Society*, 165 (1), 85–103, https://doi.org/10.1144/0016-76492006-179; B. Watts, W.S. McKerrow, and K. Richards (2005). 'Localized Quaternary uplift of south–central England.' *Journal of the Geological Society*, 162 (1), 13–24, https://doi.org/10.1144/0016-764903-127
18. Peter Japsen, Torben Bidstrup, and K. Lidmar-Bergström (2002). 'Neogene uplift and erosion of southern Scandinavia induced by the rise of the South Swedish Dome.' In: A.G. Doré, J.A. Cartwright, M.S. Stoker, J.P. Turner, and N. White (eds), *Exhumation of the North Atlantic Margin: Timing, Mechanisms and Implications for Petroleum*, London: The Geological Society, Special Publications no. 196, pp. 183–207, https://doi.org/10.1144/GSL.SP.2002.196.01.12
19. E.J. Rohling, K. Braun, K. Grant, M. Kucera, A.P. Roberts, M. Siddall, and G. Trommer (2010). 'Comparison between Holocene and Marine Isotope Stage-11 sea-level histories.' *Earth and Planetary Science Letters*, 291 (1–4), 97–105, https://doi.org/10.1016/j.epsl.2009.12.054
20. E.J. Rohling, K. Braun, K. Grant, M. Kucera, A.P. Roberts, M. Siddall, G. Trommer; Past Interglacials Working Group of PAGES (2016). 'Interglacials of the last 800,000 years.' *Reviews of Geophysics*, 54 (1), 162–219, https://doi.org/10.1002/2015RG000482
21. David L. Roberts, Panagiotis Karkanas, Zenobia Jacobs, Curtis W. Marean, and Richard G. Roberts (2012). 'Melting ice sheets 400,000 yr ago raised sea level by 13 m: past analogue for future trends.' *Earth and Planetary Science Letters*, 357–358, 226–237, https://doi.org/10.1016/j.epsl.2012.09.006; Maureen E. Raymo and Jerry X. Mitrovica (2012). 'Collapse of polar ice sheets during the stage 11 interglacial.' *Nature*, 483 (7390), 453–456, https://www.nature.com/articles/nature10891

22. D.T. Holyoak and R.C. Preece (1983). 'Evidence of a high Middle Pleistocene sea-level from estuarine deposits at Bembridge, Isle of Wight, England.' *Proceedings of the Geologists' Association*, 94 (3), 231–244, https://doi.org/10.1016/S0016-7878(83)80041-8; Preece et al., 'Pleistocene sea-level'.
23. D.A. Robinson and R.B.G. Williams (1983). 'The Sussex coast past and present.' In: The Geography Editorial Committee, University of Sussex (ed.), *Sussex: Environment, Landscape and Society*, Gloucester: Alan Sutton, pp. 50–66, http://www.sussex.ac.uk/geography/researchprojects/coastview/Introduction_background/Sussex_blue_book_coasts.pdf
24. Michael Pitts and Mark Roberts (1998). *Fairweather Eden: Life in Britain Half a Million Years Ago as Revealed by the Excavations at Boxgrove*. London: Arrow.
25. Nick Ashton, Simon G. Lewis, Isabelle De Groote, Sarah M. Duffy, Martin Bates, Richard Bates, Peter Hoare, Mark Lewis, Simon A. Parfitt, Sylvia Peglar, Craig Williams, and Chris Stringer (2014). 'Hominin footprints from Early Pleistocene deposits at Happisburgh, UK.' *PLoS One*, 9 (2): art. no. e88329, https://doi.org/10.1371/journal.pone.0088329
26. E.J. Rohling, K. Grant, Ch. Hemleben, M. Siddall, B.A.A. Hoogakker, M. Bolshaw, and M. Kucera (2008). 'High rates of sea-level rise during the last interglacial period.' *Nature Geoscience*, 1 (1), 38–42, https://www.nature.com/articles/ngeo.2007.28; A.E. Carlson (2011). 'Ice sheets and sea level in Earth's past,' *Nature Education Knowledge*, 3 (10): art. no. 3, https://www.nature.com/scitable/knowledge/library/ice-sheets-and-sea-level-in-earth-24148940
27. Victor J. Polyak, Bogdan P. Onac, Joan J. Fornós, Carling Hay, Yemane Asmerom, Jeffrey A. Dorale, Joaquín Ginés, Paola Tuccimei, and Angel Ginés (2018). 'A highly resolved record of relative sea level in the western Mediterranean Sea during the last interglacial period.' *Nature Geoscience*, 11, 860–864, https://doi.org/10.1038/s41561-018-0222-5; A. Conrad Neumann and P.J. Hearty (1996). 'Rapid sea-level changes at the close of the last interglacial (substage 5e) recorded in Bahamian island geology.' *Geology*, 24 (9), 775–778.
28. W.H. Zagwijn (1983). 'Sea-level changes in the Netherlands during the Eemian.' *Geologie en Mijnbouw*, 62, 437–450.
29. B.W. Sparks and R.G. West (1963). 'The interglacial deposits at Stutton, Suffolk.' *Proceedings of the Geologists' Association*, 74 (4), 419–432, https://doi.org/10.1016/S0016-7878(63)80001-2
30. D.H. Keen, R.S. Harmon, and J.T. Andrews (1981). 'U series and amino acid dates from Jersey.' *Nature*, 289, 162–164, https://doi.org/10.1038/289162a0

31. Andrea Dutton, Edouard Bard, Fabrizio Antonioli, Tezer M. Esat, Kurt Lambeck, and Malcolm T. McCulloch (2009). 'Phasing and amplitude of sea-level and climate change during the penultimate interglacial.' *Nature Geoscience*, 2 (5), 355–359, https://doi.org/10.1038/ngeo470
32. Andrea Dutton, F. Antonioli, and E. Bard (2009). 'A new chronology of sea level highstands for the penultimate interglacial.' *PAGES News*, 17 (2), 66–68, http://pastglobalchanges.org/download/docs/newsletter/2009-2/Special_section/science_highlights/Dutton_2009-2(66-68).pdf
33. K. Headon Davies and D.H. Keen (1985). 'The age of Pleistocene marine deposits at Portland, Dorset.' *Proceedings of the Geologists' Association*, 96 (3), 217–225, https://doi.org/10.1016/S0016-7878(85)80004-3; R.G. West, B.W. Sparks, and A.T. Sutcliffe (1960). 'Coastal interglacial deposits of the English Channel.' *Philosophical Transactions of the Royal Society B: Biological Sciences*, 243 (701), 95–133, https://doi.org/10.1098/rstb.1960.0006
34. Bethan Joan Davies, David R. Bridgland, Dave Roberts, and William E.N. Austin (2009). 'The age and stratigraphic context of the Easington Raised Beach, County Durham, UK.' *Proceedings of the Geologists' Association*, 120 (4), 183–198, https://doi.org/10.1016/j.pgeola.2009.04.001
35. James Rose, Colin A. Whiteman, Peter Allen, and Rob A. Kemp (1999). 'The Kesgrave Sands and Gravels: "pre-glacial" Quaternary deposits of the River Thames in East Anglia and the Thames Valley.' *Proceedings of the Geologists' Association*, 110 (2), 93–116, https://www.sciencedirect.com/science/article/abs/pii/S0016787899800637
36. Rob Westaway, Darrel Maddy, and David Bridgland (2002). 'Flow in the lower continental crust as a mechanism for the Quaternary uplift of south-east England: constraints from the Thames terrace record.' *Quaternary Science Reviews*, 21 (4–6), 559–603, https://www.sciencedirect.com/science/article/abs/pii/S0277379101000403
37. D.R. Bridgland and D.C. Schreve (2009). 'Implications of new Quaternary uplift models for correlation between the Middle and Upper Thames terrace sequences, UK.' *Global and Planetary Change*, 68 (4), 346–356, https://www.sciencedirect.com/science/article/pii/S0921818109000460
38. Rob Westaway, David Bridgland, and Mark White (2006). 'The Quaternary uplift history of central southern England: evidence from the terraces of the Solent River system and nearby raised beaches.' *Quaternary Science Reviews*, 25 (17–18), 2212–2250, https://www.sciencedirect.com/science/article/abs/pii/S0277379105001939; D. Maddy, D.R. Bridgland, and C.P. Green (2000). 'Crustal uplift in southern England: evidence from the river terrace records.' *Geomorphology*, 33 (3–4), 167–181, https://www.sciencedirect.com/science/article/pii/S0169555X99001208

39. Antoine et al., 'Long-term fluvial archives'.
40. Jean-Pierre Peulvast, François Bétard, and Christian Giusti (2014). 'The Seine River from Ile-de-France to Normandy: geomorphological and cultural landscapes of a large meandering valley.' In: M. Fort and M.-F. André (eds), *Landscapes and Landforms of France*, World Geomorphological Landscapes, Dordrecht: Springer Science+Business Media, pp. 17–28, https://doi.org/10.1007/978-94-007-7022-5_3; Antoine et al., 'Long-term fluvial archives'.
41. W. Wordsworth (1798). 'Lines composed a few miles above Tintern Abbey, on revisiting the banks of the Wye during a tour', 13 July.
42. Michael Roberts (1984). 'The Mosella of Ausonius: an interpretation.' *Transactions of the American Philological Association*, 114, 343–353.
43. 'London Bridge 1825 to 1967: where Thames smooth waters glide', http://www.thames.me.uk/s00049d.htm
44. Weather-history (2008). 'The Thames in London running dry...', *Netweather Community*, 31 July, https://www.netweather.tv/forum/topic/48751-the-thames-in-london-running-dry/
45. Risk Management Solutions (2007). '1607 Bristol Channel Floods, a 400 year retrospective', https://forms2.rms.com/rs/729-DJX-565/images/fl_1607_bristol_channel_floods.pdf
46. P.L. Woodworth, F.N. Teferle, R.M. Bingley, I. Shennan, and S.D.P. Williams (2009). 'Trends in UK mean sea level revisited.' *Geophysical Journal International*, 176 (1), 19–30, https://doi.org/10.1111/j.1365-246X.2008.03942.x
47. R.L. Barnett et al. (2020). 'Nonlinear landscape and cultural response to sea-level rise.' *Science Advances*, 6 (45).
48. Julia Stockamp, Zhenhong Li, Paul Bishop, Jim Hansom, Alistair Rennie, Elizabeth Petrie, Akiko Tanaka, Richard Bingley, and Dionne Hansen (2015). 'Investigating glacial isostatic adjustment in Scotland with Insar and GPS observations.' In: *Proceedings of the 'Fringe 2015 Workshop'*, Frascati, Italy, 23–27 March (ESA SP-731, May 2015), http://proceedings.esa.int/files/171.pdf
49. W. Ritchie (1966). 'The post-glacial rise in sea-level and coastal changes in the Uists.' *Transactions of the Institute of British Geographers*, 39, 79–86, https://doi.org/10.2307/621677

CHAPTER 11

1. R.M.W. Musson (2008). 'The seismicity of the British Isles to 1600', British Geological Survey Open Report, OR/08/049, http://www.earthquakes.bgs.ac.uk/historical/data/studies/MUSS008/MUSS008.pdf
2. J.W. Evans (1971). *St Davids Cathedral*. Andover: Pitkin.

3. A. Grove (1981). The great earthquake of 1382. *Journal of Kent Local History*, 9–10.
4. S.L. Sargeant, P.J. Stafford, R. Lawley, G. Weatherill, A.-J.S. Weston, J.J. Bommer, P.W. Burton, M. Free, R.M.W. Musson, T. Kuuyuor, and T. Rossetto (2008). 'Observations from the Folkestone, UK, earthquake of 28 April 2007.' *Seismological Research Letters*, 79 (5), 672–687, http://peterjstafford.com/papers/journals/Sargeant%20et%20al%20(2008).pdf
5. Macroseismology (2016). *Royal Observatory of Belgium*, http://seismologie.oma.be/en/research/seismology/macroseismology
6. L. Hickman (2013). 'Why north Wales is the UK's earthquake hotspot', *The Guardian*, 29 May, https://www.theguardian.com/world/shortcuts/2013/may/29/llyn-peninsula-uk-earthquake-hotspot
7. M.E.A. Ritchie, R.M.W. Musson, and N.H. Woodcock (1990). 'The Bishop's Castle (UK) earthquake of 2 April 1990.' *Terra Nova*, 2 (4), 390–400, https://doi.org/10.1111/j.1365-3121.1990.tb00091.x
8. Brian Baptie, Lars Ottemoller, Susanne Sargeant, Glenn Ford, and Aoife O'Mongain (2005). 'The Dudley earthquake of 2002: a moderate sized earthquake in the UK.' *Tectonophysics*, 401 (1–2), 1–22, https://doi.org/10.1016/j.tecto.2005.02.010
9. L. Ottemöller, B. Baptie, and N.J.P. Smith (2009). 'Source parameters for the 28 April 2007 Mw 4.0 earthquake in Folkestone, United Kingdom.' *Bulletin of the Seismological Society of America*, 99 (3), 1853–1867, https://doi.org/10.1785/0120080244
10. B. Baptie (2010). 'Seismogenesis and state of stress in the UK', http://nora.nerc.ac.uk/id/eprint/9574/1/baptie_tectonophysics_2010_preprint.pdf
11. Roger M.W. Musson (2004). 'A critical history of British earthquakes.' *Annals of Geophysics*, 47 (2/3), 597–609, https://www.annalsofgeophysics.eu/index.php/annals/article/view/3325
12. Rob Westaway (2006). 'Seismic imaging of a hot upwelling beneath the British Isles: comment.' *Geology*, 34 (1): art. no. e95, https://doi.org/10.1130/G22093.1
13. D.T. Aldiss, H.F. Burke, B. Chacksfield, R. Bingley, N. Teferle, S. Williams, D. Blackman, R. Burren, and N. Press (2014). 'Geological interpretation of current subsidence and uplift in the London area, UK, as shown by high precision satellite based surveying.' *Proceedings of the Geologists' Association*, 125 (1), 1–13, https://doi.org/10.1016/j.pgeola.2013.07.003
14. D.T. Aldiss (2013). 'Under-representation of faults on geological maps of the London region: reasons, consequences and solutions.' *Proceedings of the Geologists' Association*, 124 (6), 929–945, http://nora.nerc.ac.uk/id/eprint/503233/1/Aldiss_London_faulting_V3.pdf

15. J.R. Lee, R. Haslam, M.A. Woods, J. Rose, R.L. Graham, J.R. Ford, D.I. Schofield, T.I. Kearsey, and C.N. Williams (2020). 'Plio-Pleistocene fault reactivation within the Crag Basin, eastern UK: implications for structural controls of landscape development within an intraplate setting.' *Boreas*, 49 (4), 685–708, https://doi.org/10.1111/bor.12462
16. P.L. Gibbard, J.A. Zalasiewicz, and S.J. Mathers (1998). 'Stratigraphy of the marine Plio-Pleistocene Crag deposits of East Anglia.' In: P.L. Gibbard and T. van Kolfschoten (eds), *The Dawn of the Quaternary*, Haarlem: Mededelingen Netherlands Instituut voor Toegepaste Goewetenschappen TNO 60, pp. 239–262; Stéphane Baize, Michel Coulon, Christian Hibsch, Marc Cushing, Francis Lemeille, and Erwan Hamard (2007). 'Non-tectonic deformations of Pleistocene sediments in the eastern Paris basin, France.' *Bulletin de la Société Géologique de France*, 178 (5), 367–381, https://doi.org/10.2113/gssgfbull.178.5.367
17. S. Holloway and R.A. Chadwick (1986). 'The Sticklepath-Lustleigh fault zone: Tertiary sinistral reactivation of a Variscan dextral strike-slip fault.' *Journal of the Geological Society*, 143 (3), 447–452, https://jgs.lyellcollection.org/content/143/3/447
18. R.W. Gallois (2006). 'The geology of the hot springs at Bath Spa, Somerset.' *Geoscience in South-West England*, 11 (3), 168–173, http://nora.nerc.ac.uk/id/eprint/4841/1/Hot_Springs_2006.pdf; R.W. Gallois (2007). 'The formation of the hot springs at Bath Spa, UK.' *Geological Magazine*, 144 (4), 741–747, http://nora.nerc.ac.uk/id/eprint/507196/1/Hot%20springs%201.pdf
19. G.A. Kellaway (1993). 'The hot springs of Bristol and Bath.' *Proceedings of the Ussher Society*, 8, 83–88, https://ussher.org.uk/wp-content/uploads/journal/1993/01-Kellaway_1993.pdf
20. W.G. Burgess, J.H. Black, and A.J. Cook (1991). 'Regional hydrodynamic influences on the Bath-Bristol springs.' In: G.A. Kellaway (ed.), *Hot Springs of Bath*, Bath: Bath City Council, pp. 171–177.
21. G.A. Kellaway (1994). 'Environmental factors and the development of Bath Spa, England.' *Environmental Geology*, 24, 99–111, https://doi.org/10.1007/BF00767883
22. Nick Catford (2015). 'Shakespeare Colliery', *Subterranea Britannica*, https://www.subbrit.org.uk/sites/shakespeare-colliery/
23. R.H. Bird (1976). 'Notes on subterranean temperatures in metal mines.' *British Mining*, 3, 16–20.
24. Luciane Licour (2014). 'The geothermal reservoir of Hainaut: the result of thermal convection in a carbonate and sulfate aquifer.' *Geologica Belgica*, 17 (1), 75–81, https://popups.uliege.be/1374-8505/index.php?id=4411&file=1&pid=4401

25. F.C. Brassington (2007). 'A proposed conceptual model for the genesis of the Derbyshire thermal springs.' *Quarterly Journal of Engineering Geology and Hydrogeology*, 40 (1), 35–46, https://doi.org/10.1144/1470-9236/05-046
26. John Gunn, Simon H. Bottrell, David J. Lowe, and Stephen R.H. Worthington (2006). 'Deep groundwater flow and geochemical processes in limestone aquifers: evidence from thermal waters in Derbyshire, England, UK.' *Hydrogeology Journal*, 14 (6), 868–881, https://doi.org/10.1007/s10040-006-0022-7
27. BGS Earthquake Seismology (2012). 'Manchester Earthquakes Sequence, October–November 2002', British Geological Survey, http://www.earthquakes.bgs.ac.uk/research/manchester_sequence.html
28. 'Notes on individual earthquakes', British Geological Survey, https://web.archive.org/web/20070613072349/http://www.quakes.bgs.ac.uk/earthquakes/historical/historical_listing.htm
29. M.J.M. Cunningham, A.W.E. Phillips, and A.L. Densmore (2004). 'Evidence for Cenozoic tectonic deformation in SE Ireland and near off shore.' *Tectonics*, 23 (6), TC6002, https://doi.org/10.1029/2003TC001597
30. Alan Judd, Peter Croker, Louise Tizzard, and Carolyn Voisey (2007). 'Extensive methane-derived authigenic carbonates in the Irish Sea.' *Geo-Marine Letters*, 27, 259–267, https://doi.org/10.1007/s00367-007-0079-x
31. E. Serpelloni, G. Vannucci, S. Pondrelli, A. Argnani, G. Casula, M. Anzidei, P. Baldi, and P. Gasperini (2007). 'Kinematics of the Western Africa–Eurasia plate boundary from focal mechanisms and GPS data.' *Geophysical Journal International*, 169 (3), 1180–1200, https://doi.org/10.1111/j.1365-246X.2007.03367.x
32. Maria Ana Baptista, Jorge Miguel Miranda, Josep Batlló, Filipe Lisboa, Joaquim Luis, and Ramon Maciá (2016). 'New study on the 1941 Gloria Fault earthquake and tsunami.' *Natural Hazards and Earth System Sciences*, 16 (8), 1967–1977, https://doi.org/10.5194/nhess-16-1967-2016
33. J. Galindo-Zaldívar, A. Maldonado, and A.A. Schreider (2003). 'Gorringe Ridge gravity and magnetic anomalies are compatible with thrusting at a crustal scale.' *Geophysical Journal International*, 153 (3), 586–594, https://doi.org/10.1046/j.1365-246X.2003.01922.x
34. N. Zitellini, E. Gràcia, L. Matias, P. Terrinha, M.A. Abreu, G. DeAlteriis, J.P. Henriet, J.J. Dañobeitia, D.G. Masson, T. Mulder, R. Ramella, L. Somoza, and S. Diez (2009). 'The quest for the Africa–Eurasia plate boundary west of the Strait of Gibraltar.' *Earth and Planetary Letters*, 280 (1–4), 13–50, https://doi.org/10.1016/j.epsl.2008.12.005
35. Y. Fukao (1973). 'Thrust faulting at a lithospheric plate boundary: the Portugal earthquake of 1969.' *Earth and Planetary Science Letters*, 18 (2), 205–216, https://doi.org/10.1016/0012-821X(73)90058-7

36. *London Evening News*, 6–9 December 1755.
37. *Scots Magazine*, 1755.
38. R. Muir-Wood and A. Mignan (2009). 'A phenomenological reconstruction of the Mw9 Nov 1st 1755 earthquake source.' In: Luiz A. Mendes-Victor, Carlos Sousa Oliveira, João Azevedo, and António Ribeiro (eds), *The 1755 Lisbon Earthquake: Revisited*, Geotechnical, Geological and Earthquake Engineering, 7, Dordrecht: Springer, pp. 121–146, https://doi.org/10.1007/978-1-4020-8609-0_8
39. M.A. Baptista, J.M. Miranda, F. Chierici, and N. Zitellini (2003). 'New study of the 1755 earthquake source based on multi-channel seismic survey data and tsunami modeling.' *Natural Hazards and Earth System Sciences*, 3, 333–340.
40. Letters sent from Lisbon to Joseph Salvador and from Mr Molloy to Keane Fitzgerald FRS. 'An account of the earthquake at Lisbon 31st March, 1761.' *Philosophical Transactions of the Royal Society of London*, 52 (1761–1762), presented 23 April 1761, pp. 141–143.
41. Diario Noticias Gazeta de Lisboa 1761, pp. 380 and 386.
42. By Thomas Heberden, MDFRS. Communicated by William Heberden, MDFRS (1761). 'An account of the earthquake felt in the island of Madeira, March 31, 1761.' *Philosophical Transactions of the Royal Society of London*, 52 (1761–1762), 155–156.
43. W. Borlase (1761). 'Some account of the extraordinary agitation of the waters in Mount's Bay and other places on the 31st of March 1761.' *Philosophical Transactions of the Royal Society of London*, 52 (1761–1762), 418–453.
44. Paul Simon (2017). 'Weathereye', *The London Times*, 19 December.
45. Borlase, 'Some account of the extraordinary agitation.'
46. J. Murillas, D.G. Mougenot, G. Boillot, M.C. Comas, E. Banda, and A. Mauffret (1990). 'Structure and evolution of the Galicia Interior Basin (Atlantic western Iberian continental margin).' *Tectonophysics*, 184 (3–4), 297–319, https://doi.org/10.1016/0040-1951(90)90445-E; M. Druet, A. Muñoz-Martín, J.L. Granja-Bruña, A. Carbó-Gorosabel, J. Acosta, P. Llanes, and G. Ercilla (2018). 'Crustal structure and continent-ocean boundary along the Galicia continental margin (NW Iberia): insights from combined gravity and seismic interpretation.' *Tectonics*, 37 (5), 1576–1604, https://doi.org/10.1029/2017TC004903
47. Bernd Andeweg (2002). 'Cenozoic tectonic evolution of the Iberian Peninsula: effects and causes of changing stress fields', PhD thesis, de Vrije Universiteit Amsterdam.
48. P. Lacan and M. Ortuño (2012). 'Active tectonics of the Pyrenees: a review.' *Journal of Iberian Geology*, 38 (1), 9–30, https://doi.org/10.5209/rev_JIGE.2012.v38.n1.39203

49. J. Henning Illies (1981). 'Mechanism of graben formation.' *Developments in Geotectonics*, 17, 249–266, https://doi.org/10.1016/B978-0-444-41956-9.50022-8

CHAPTER 12

1. B. Mandelbrot (1967). 'How long is the coast of Britain? Statistical self-similarity and fractional dimension.' *Science*, 156 (3775), 636–638, https://doi.org/10.1126/science.156.3775.636
2. Geoffrey King (1983). 'The accommodation of large strains in the upper lithosphere of the earth and other solids by self-similar fault systems: the geometrical origin of b-value.' *Pure and Applied Geophysics*, 121 (5/6), 761–815, https://doi.org/10.1007/BF02590182
3. Charles Robert Darwin (1859). *On the Origin of Species*, first edition, p. 286.
4. Henry Faul (1978). 'A history of geologic time: "it is perhaps a little indelicate to ask of our Mother Earth her age..." (Arthur Holmes 1913).' *American Scientist*, 66 (2), 159–165.
5. Martin D. Hurst, Dylan H. Rood, Michael A. Ellis, Robert S. Anderson, and Uwe Dornbusch (2016). 'Recent acceleration in coastal cliff retreat rates on the south coast of Great Britain.' *Proceedings of the National Academy of Sciences*, 113 (47), 13336–13341, https://doi.org/10.1073/pnas.16130441
6. *The Independent on Sunday* (1993). 'A town under siege by the sea', 17 October.
7. D.A. Robinson and R.B.G. Williams (1983). 'The Sussex coast past and present.' In: The Geography Editorial Committee, University of Sussex (ed.), *Sussex: Environment, Landscape and Society*, Gloucester: Alan Sutton, pp. 50–66, http://www.sussex.ac.uk/geography/researchprojects/coast-view/Introduction_background/Sussex_blue_book_coasts.pdf
8. Environment Agency (2016). 'Managing flood and coastal erosion in England 1 April 2015 to 31 March 2016'. p. 32, https://assets.publishing.service.gov.uk/media/5a8217d240f0b62305b92714/National_Flood_Risk_Report_LIT_10517.pdf
9. Nicholas Schoon (1993). 'Up to six metres of coastline claimed by the sea each year: some coastal defences can cause erosion on nearby beaches, scientists warn', *The Independent*, 8 June, https://www.independent.co.uk/news/uk/up-to-six-metres-of-coastline-claimed-by-the-sea-each-year-some-coastal-defences-can-cause-erosion-1490451.html
10. J. Eddison, A.P. Carr, and I.P. Jolliffe (1983). 'Endangered coastlines of geomorphological importance.' *The Geographical Journal*, 149 (1), 39–75, https://doi.org/10.2307/633341

11. M.E. Shaw Champion, N.J. White, S.M. Jones, and J.P.B. Lovell (2008). 'Quantifying transient mantle convective uplift: an example from the Faroe–Shetland basin.' *Tectonics*, 27 (1), https://doi.org/10.1029/2007TC002106
12. G.F. Mitchell (1980). 'The search for Tertiary Ireland.' *Journal of Earth Sciences, Royal Dublin Society*, 3 (1), 13–33.
13. Chapter XIV 'Central Chile' of Charles Darwin's 1839 *Voyage of the Beagle*, originally published as *Journal and Remarks*.
14. Bert Hansen (1970). 'The early history of glacial theory in British geology.' *Journal of Glaciology*, 9 (55), 135–141, https://doi.org/10.3189/S0022143000026861
15. Joshua Trimmer (1831). 'On the diluvial deposits of Caernarvonshire, between the Snowdon chain of hills and the Menai strait, and on the discovery of marine shells in diluvial sand and gravel on the summit of Moel Tryfane, near Caernarvon, 1000 ft above the level of the sea. [Read 8 June 1831.]' *Proceedings of the Geological Society of London*, 1 (1826–33), 331–332. [Vols. 4, 9, 11.]
16. C.R. Darwin (1842). 'Notes on the effects produced by the ancient glaciers of Caernarvonshire, and on the boulders transported by floating ice', *The London, Edinburgh and Dublin Philosophical Magazine*, 21 (September), 180–188, http://darwin-online.org.uk/content/frameset?itemID=F1660&viewtype=text&pageseq=1
17. Letter from C.R. Darwin to T.F. Jamieson, 6 September 1861, Darwin Correspondence Project, https://www.darwinproject.ac.uk/letter/DCP-LETT-3247.xml
18. P.J. Boylan (1998). 'Lyell and the dilemma of Quaternary glaciation.' In: D.J. Blundell and A.C. Scott (eds), *Lyell: The Past is the Key to the Present*, London: The Geological Society, Special Publications no. 143, pp. 145–159, https://sp.lyellcollection.org/content/specpubgsl/143/1/145.full.pdf
19. P. Walsh, M. Boulter, and I. Morawiecka (1999). 'Chattian and Miocene elements in the modern landscape of western Britain and Ireland.' In: B.J. Smith, W.B. Whalley, and P. Warke (eds), *Uplift, Erosion and Stability: Perspectives on Long-Term Landscape Development*, London: The Geological Society, Special Publications no. 162, pp. 45–63, https://doi.org/10.1144/GSL.SP.1999.162.01.04
20. G.L.H. Davies and N. Stephens (1978). *The Geomorphology of the British Isles: Ireland*, London: Methuen.
21. Jean-Louis Lagarde, Stéphane Baize, and Daniel Amorese (2000). 'Active tectonics, seismicity and geomorphology with special reference to Normandy (France).' *Journal of Quaternary Science*, 15 (7), 745–758,

https://doi.org/10.1002/1099-1417(200010)15:7<745::AID-JQS534>3.0.CO;2-6; Jean-Pierre Larue (2005). 'The status of ravine-like incisions in the dry valleys of the Pays de Thelle (Paris basin, France).' *Geomorphology*, 68 (3–4), 242–256, https://doi.org/10.1016/j.geomorph.2004.11.018

22. S. Bonnet (1998). 'Tectonique et dynamique du relief: le socle armoricain au Pléistocène', thesis, University of Rennes I.
23. R. Muir-Wood (1985). *The Dark Side of the Earth: The Battle for the Earth Sciences, 1800–1980*, London: Allen & Unwin.
24. Søren B. Nielsen, Randell Stephenson, and Erik Thomsen (2008). 'Dynamics of Mid-Palaeocene North Atlantic rifting linked with European intra-plate deformations.' *Nature*, 450 (7172), 1071–1074, https://doi.org/10.1038/nature06379
25. Jan de Jager (2003). 'Inverted basins in the Netherlands, similarities and differences.' *Geologie en Mijnbouw*, 82 (4), 339–349, https://doi.org/10.1017/S0016774600020175
26. Kevin H. Smith, Patrick Whitley, Geoffrey S. Kimbell, Martin Kubala, and H. Johnson (2009). 'Enhancing the prospectivity of the Wyville Thomson Ridge.' In: *2nd Faroe Islands Exploration Conference: Annales Societas Scientiarum Faeroensis*, Tórshavn: Faroese Academy of Sciences, pp. 286–302.

CHAPTER 13

1. A.K. Quirie, N. Schofield, A. Hartley, M.J. Hole, S.G. Archer, J.R. Underhill, D. Watson, and S.P. Holford (2019). 'The Rattray Volcanics: Middle Jurassic fissure volcanism in the UK central North Sea.' *Journal of the Geological Society*, 176 (3), 462–481, https://doi.org/10.1144/jgs2018-151
2. 'Fullers Earth', *The Woburn Sands Collection*, https://www.mkheritage.org.uk/wsc/environment/fullers-earth/
3. Frederick Reinig, Lukas Wacker, Olaf Jöris, Clive Oppenheimer, Giulia Guidobaldi, Daniel Nievergelt, Florian Adolphi, Paolo Cherubini, Stefan Engels, Jan Esper, Alexander Land, Christine Lane, Hardy Pfanz, Sabine Remmele, Michael Sigl, Adam Sookdeo, and Ulf Büntgen (2021). 'Precise date for the Laacher See eruption synchronizes the Younger Dryas.' *Nature*, 595 (7865), 66–69, https://doi.org/10.1038/s41586-021-03608-x; P. Cherubini, F. Reinig, and J. Esper (2021). 'Eruption of the Laacher See Volcano redated', *WSL*, 30 June, https://www.wsl.ch/en/news/2021/06/eruption-of-the-laacher-see-volcano-redated.html

4. Sibrecht Reniere, Roland Dreesen, Gilles Fronteau, Tatjana Gluhak, Eric Goemaere, Else Hartoch, Paul Picavet, and Wim De Clercq (2016). 'Querns and mills during Roman times at the northern frontier of the Roman Empire (Belgium, northern France, southern Netherlands, western Germany): unravelling geological and geographical provenances, a multidisciplinary research project.' *Lithic Studies*, 3 (3), 403–428, https://doi.org/10.2218/jls.v3i3.1640
5. Michael Baales, Olaf Jöris, Martin Street, Felix Bittmann, Bernhard Weninger, and Julian Wiethold (2002). 'Impact of the late glacial eruption of the Laacher See Volcano, central Rhineland, Germany.' *Quaternary Research*, 58 (3), 273–288, https://doi.org/10.1006/qres.2002.2379
6. Hans-Ulrich Schmincke, Cornelia Park, and Eduard Harms (1999). 'Evolution and environmental impacts of the eruption of Laacher See Volcano (Germany) 12,900 a BP.' *Quaternary International*, 61 (1), 61–72, https://doi.org/10.1016/S1040-6182(99)00017-8
7. Felix Riede (2008). 'The Laacher See-eruption (12,920 BP) and material culture change at the end of the Allerød in northern Europe.' *Journal of Archaeological Science*, 35 (3), 591–599, https://doi.org/10.1016/j.jas.2007.05.007
8. Corné Kreemer, Geoffrey Blewitt, and Paul M. Davis (2020). 'Geodetic evidence for a buoyant mantle plume beneath the Eifel volcanic area, NW Europe.' *Geophysical Journal International*, 222 (2), 1316–1332, https://doi.org/10.1093/gji/ggaa227
9. J.W. Faithfull, M.J. Timmerman, B.G.J. Upton, and M.S. Rumsey (2012). 'Mid-Eocene renewal of magmatism in NW Scotland: the Loch Roag Dyke, Outer Hebrides.' *Journal of the Geological Society*, 169 (2), 115–118, https://doi.org/10.1144/0016-76492011-117
10. Martin H.P. Bott and Jacqueline D.J. Bott (2004). 'The Cenozoic uplift and earthquake belt of mainland Britain as a response to an underlying hot, low-density upper mantle.' *Journal of the Geological Society*, 161 (1), 19–29, https://jgs.lyellcollection.org/content/161/1/19
11. Saskia Goes, Wim Spakman, and Harmen Bijwaard (1999). 'A lower mantle source for central European volcanism.' *Science*, 286 (5446), 1928–1931, https://doi.org/10.1126/science.286.5446.1928

Index

For the benefit of digital users, indexed terms that span two pages (e.g., 52–53) may, on occasion, appear on only one of those pages.